Advances in Superplasticity
AND
Superplastic Forming

Advances in Superplasticity
AND
Superplastic Forming

Proceedings of a Symposium sponsored by
the MDMD Shaping and Forming Committee,
held at the TMS/ASM Materials Week
2 - 5 November, 1992

Edited by

N. Chandra
H. Garmestani
R.E. Goforth

A Publication of

Minerals • Metals • Materials

A Publication of The Minerals, Metals & Materials Society
420 Commonwealth Drive
Warrendale, Pennsylvania 15086
(412) 776-9000

Printed in the United States of America
Library of Congress Catalog Number 93-79176
ISBN Number 0-87339-250-7

If you are interested in purchasing a copy of this book, or if you
would like to receive the latest TMS publications catalog, please
telephone 1-800-759-4867.

Foreword

A symposium on Advances in Superplasticity and Superplastic forming was held in TMS Chicago Fall meeting during November 2-5, 1992. The meeting was organized by Professors. N. Chandra, H. Garmestani and R.E. Goforth and was extremely well attended. There were three sessions with 15 papers presented, with renewed interest in superplasticity. The interest ranged from the understanding of the fundamental deformation mechanisms to the application of this phenomenon to new materials including metal matrix composites and ceramics. The symposium proceedings will be published by TMS and should be available to the readers by mid-April 93.

Constitutive Equations

The simplest form of constitutive equation suggested for superplastic material is the power-law model of the type $\sigma = K\dot{\epsilon}^m$. The strain-rate sensitivity m is believed to be an indicator of the maximum ductility in the material and hence a process cycle to maintain the maximum value of m is used in industry. This corresponds to a constant value of true $\dot{\epsilon}$. Hamilton and coworkers reexamined this hypothesis for the conventional Ti-6Al-4V material and have suggested an optimized deformation path that takes into account the grain growth kinetics. It is shown that the total forming time can be reduced by following a strain-rate path based on geometric instability which depends on not only m but also the grain growth exponent p.

Srinivasan extended the instability analysis of Hamilton from statically recrystallized materials, e.g., Ti-6Al-4V, Al 7475 to dynamically recrystallizing materials like Al-Li alloys. He used extensive mechanical test results of three materials in this group with varying geometric and processing parameters to indicate that the total elongation is affected by microstructural dynamics and not merely by m. Goforth attempted to describe the superplastic flow of dynamically recrystallizing Al-Li alloys using Ashby-Verral and two other models. The experimental observation could not be explained using the value of a constant B as proposed in the original model; however, when this constant was manipulated a decent correlation was noticed. He argued that the pure creep activity (diffusional and dislocational) may not be adequate and that dislocation slip, especially in the subcells contribute to the overall strain-rate in this class of materials. Leuser and coworkers proposed a model that explicitly takes into account the effect of grain growth in the superplastic behavior of materials by experimentally investigating a quasi single phase metal (coronze-638) and a microduplex (ultra-high carbon steel) material. A separate grain growth equation has been proposed which depends on strain and strain-rate; thus different distributions of grain-size with the same mean value lead to a different mechanical effects. The grain-growth laws are incorporated in the constitutive equations for grain boundary sliding and dislocational creep constitutive equations.

Superplasticity in Ceramics and Composites and High strain-rate Superplasticity

One of the technological limitations on the widespread industrial use of superplastic forming is the slow strain-rate of the order of 10^{-4} per second of deformation. Efforts are underway to increase the $\dot{\epsilon}$ at which superplasticity is observed, typically referred to as the high strain-rate superplasticity. Also hard to form materials using conventional methods like intermetallics and ceramics exhibit superplasticity spurring considerable research. The topical review was provided by Nieh and Wadsworth who showed a series of Al based alloys (Al 2024, 2124, 6061 and 7475) with addition of particulates and whiskers of Silicon Carbide and Silicon Nitride exhibited superplastic behavior at high strain-rates. The detailed article appeared in Journal of Metals, November 1992 issue and is hence not included in the proceedings. Though there is no conclusive evidence in explaining the origin of High strain-rate superplasticity, the authors believe that a liquid phase in the interface between the matrix and reinforcement plays a major role.

Pilling and coworkers studied the superplastic behavior of Al_2O_3–ZrO_2–Al_2TiO_5 ceramic in compression, and reported that superplasticity was observed upto a strain-rate of 2×10^{-2} per sec, with a relatively high flow stress of 125 MPa. He observed that maintaining fine grain size is very critical and in this ceramic, ternary blends resisted the grain growth. In the next paper superplasticity of an intermetallic material Nickel Silicide (modified with the addition of V and Mo for increasing the grain stability) was presented by Stoner and Mukerjee. This intermetallic material is potential candidate for high temperature application and is shown to exhibit superplasticity at 1343^0K with $\dot{\epsilon}$ ranging from 10^{-2} to 10^{-2} per sec. Total elongation depended strongly on the microstructure, and interestingly transverse ductility to failure was found to be more than that in the longitudinal direction. Though the rate of cavity nucleation in the transverse direction was more, the rate of cavity linkage (coalescence) is less, contributing to the anomalous behavior. Another intermetallic compound actively considered in high temperature applications both in monolithic and composite form is the near-gamma Titanium Aluminide (Ti-48Al-2Cr-2NB-1Ta or Mo). Bampton investigated the superplastic behavior of this material at three different temperature (1037^0C, 1093^0C and 1204^0C), above and below the eutectoid temperature between α and α_2. A thorough investigation of microstructure, step-strain-rate tests, cavitation studies indicate that the alloy as-processed to produce the duplex microstructure of equiaxed γ and lamellar colonies of γ and α_2 is not suitable for superplastic forming.

Novel Experimental Techniques and Process Modeling

The microstructural prerequisites for superplasticity are fine, equiaxed and stable grains, presence of second phase with compatible strength, size and distribution; these materials exhibit large elongation when deformed within a narrow range of temperatures and strain-rates. The physical mechanisms that are believed to be active are grain-boundary sliding with grain compatibility achieved through diffusional flow or dislocational slip. There have been many theories proposed to explain the origin of superplasticity and the role of different deformation mechanisms that collectively contribute to the overall elongations observed in superplastic materials. Unfortunately, the actual experimental measurements of the role of the different deformation mechanisms have relied upon optical and electron microscopy works which are at best qualitative in nature. There are a couple of research work in this symposium that uses texture measurements using X-ray and a novel Electron Backscattered Kikuchi Diffraction method for this purpose.

Garmestani, Chandra and coworkers studied the evolution of texture in Ti-6Al-4V when the material was superplastically stretched in a state of balanced biaxial tension. The novelty of the analysis lies in using Electron Backscattered Kikuchi Diffraction technique in obtaining the complete pole figures of the α phase. This method can also be used to get orientation distribution function and the distribution of misorientation angles between adjacent grains. A special problem faced in this material was that though Ti-6Al-4V has α to β ratio of 90% to 10% at room temperature (where the texture is measured), at the superplastic temperature of 927^0C, the ratio significantly changes to 50–50%. Hence the texture results of the 90% α phase actually consists of approximately 40% β which has a phase transformation during the cooling subsequent to the mechanical deformation process. They observed from the pole figures and orientation distribution functions that there is a loss of texture with strain upto a certain value of biaxial strain indicating that grain boundary sliding was the primary mechanism of deformation. Bieler used texture analysis using x-ray diffraction method to study the effect of different superplastic deformation mechanisms, grain-boundary sliding, dislocation slip, diffusional creep, and recrystallization in the overall behavior of high strain-rate superplasticity seen in mechanically alloyed IN90211. Since grain boundary sliding weakens existing texture components and, dislocation slip and dynamic recrystallization create specific texture creep, Bieler used the orientation distribution functions of specimens deformed to different strain levels to obtain the specific deformation mechanisms.

Chandra and Rama presented the superplastic process modeling of a complex 3-dimensional component using finite element method. In general, components of complex 3-dimensional shape offer the economical edge when formed using superplastic method over other conventional forming methods; thus modeling the forming process of such complex shapes becomes very important. The success of forming operation relies heavily on maintaining the optimum equivalent strain-rate at all times of forming process. Chandra presented the details of a three-dimensional finite element formulation using a new element that uses significantly less computational effort than other types of elements. Deformation profiles, surface plots of thickness and strain-rate during the entire forming cycle of complex 3-D part made of Al-Li alloy were presented. The model predicted the process parameters of pressure-time cycle which was used in the actual forming process. Good comparisons were observed between the experimentally measured thickness distributions and the numerical results.

Hartley et. al. examined the optimal processing conditions of an Al-Li alloy (Weldalite 049) to not only maintain the optimum superplastic conditions during forming but also obtain an acceptable post-spf properties when combined with an heat treatment cycle. This additional requirement in the Al-Li material is necessitated by the lack of strengthening response from a pure ageing process. A series of microstructural and mechanical property evaluation on specimens with different post-spf heat treatment procedures were used to identify the optimum overall process cycle.

The editors would like to thank the authors for their cooperation in the arduous task of manuscript preparation. Also, we acknowledge the support and assistance provided by the Shaping and Forming committee of TMS in this endeavor.

N. Chandra
FAMU/FSU College of Engineering
Tallahassee, Florida

R.E. Goforth
Texax A&M University
College Station, TX

H. Garmestani
FAMU/FSU College of Engineering
Tallahassee, Florida

Table of Contents

CONSTITUTIVE EQUATIONS

SUPERPLASTICITY IN COMPOSITES
AND HIGH STRAIN-RATE SUPERPLASTICITY

NOVEL EXPERIMENTAL TECHNIQUES AND PROCESS MODELING

CONSTITUTIVE EQUATIONS

Designing Optimized Deformation Paths

for Superplastic Ti-6Al-4V

C. H. Johnson, C. H. Hamilton, H. M. Zbib, S. K. Richter
Department of Mechanical and Materials Engineering
Washington State University, Pullman, WA 99164-2920

Abstract

The mechanistic modeling of the superplastic deformation process has resulted in general relations in which the strain rate can be related to the applied stress. The form of these relations have been adopted here for a micro-duplex Ti alloy (Ti-6Al-4V) and coupled with recent research from which the time- and deformation-dependent grain coarsening kinetics can be described. The grain size is well known to strongly influence the flow properties, and grain coarsening can cause hardening as well as a decrease in the Region II-Region III transition strain rate. These combined effects are important and make difficult the task of describing and predicting stability of superplastic deformation in the tensile test and in superplastic forming. In this study, parametric values for deformation and grain coarsening were established under constant (average) strain rate conditions, and utilized to predict corresponding behavior under other deformation conditions. Concepts for optimizing deformation paths were explored through numerical modeling and examined experimentally. It is shown that there is little opportunity to delay the onset of non-uniform strain in the tensile tests due to non-uniform stress states, but that subsequent necking can be limited by control of the post-uniform deformation process. A concept is proposed and tested experimentally for designing a variable strain rate path suitable for minimizing non-uniform deformation while deforming as rapidly as possible.

Advances in Superplasticity and Superplastic Forming
Edited by N. Chandra, H. Garmestani, R.E. Goforth
The Minerals, Metals & Materials Society, 1993

Introduction

Superplastic forming (SPF) is a process being used increasingly for forming a range of materials for both aerospace and non-aerospace applications. As the process finds increasing use, there is a concurrent interest in increasing productivity, especially through more rapid forming since SPF is typically a slow process, requiring forming times of tens of minutes to several hours in some cases. The problem encountered is that rapid localization and failure occur at the higher strain rates, and superplastic deformation (SPD) must therefore be restricted in terms of strain rate. There has been success demonstrated in increasing superplastic strain rates through material design, and in some research materials very high forming rates have been reported [1]. However, it is nonetheless desirable to maximize forming rates for the more conventional, commercially available superplastic forming alloys. In this case, process design is believed to be one way to improve productivity with these materials.

Forming times are usually based on gas pressurization profiles designed to impose an approximately constant strain rate in a selected region of a part. However, it is found that superplastic ductility is dependent on the strain rate path; and, for example, quite different elongations are often found for different strain rate paths for the same initial strain rate [2]. Recent research has also shown that the strain hardening in superplastic deformation can play an important role in the necking resistance of superplastic materials [3]. Recognition of strain hardening in an Al-Li alloy has lead to design of a strain rate path based on experimentally measured strain and strain-rate parameters which demonstrated significant reduction in time for deforming to a relatively high superplastic strain relative to the constant strain rate. While the kind of microstructural evolution is different, it is nonetheless considered that hardening in a Ti-6Al-4V alloy, due to grain growth, can result in a similar effect and potential for more rapid forming if an appropriate deformation path can be established and used. In the study reported herein, a concept is evaluated in the uniaxial tension test whereby the deformation rate is established through an analytical process design in which the strain, or flow, hardening is considered as well as the strain-rate hardening through use of a microstructure-based constitutive relation which incorporates grain growth kinetics.

Experimental

Constitutive Relation

Ti-6Al-4V exhibits high elongation at temperatures about 850 to 925C and is quite strain rate sensitive. Typical of fine grain superplasticity, the stress/strain rate behavior exhibits Regions I, II, and III (threshold, SPD, and power law creep) [4]. All three regions are represented in various superposed constitutive relations [5, 6, 7] in which the strain rates due to SPD and creep mechanisms are assumed to be simultaneous:

$$\dot{\varepsilon} = \dot{\varepsilon}_s + \dot{\varepsilon}_c \qquad (1)$$

where $\dot{\varepsilon}$ is the strain rate, $\dot{\varepsilon}_s$ is the superplastic component and $\dot{\varepsilon}_c$ is the creep component of the sum. Both components are generally functions of stress, temperature, diffusivity, and material parameters. The SPD term is dependent on microstructure, as opposed to the creep term which is not.

By constraining our test conditions for this research, we can simplify the relation and focus on the time-minimizing aspect of the process. Specifically, the test temperature has been fixed at 900C. For further simplification, an assumption concerning the diffusion mechanism must be made. At 900C, grain boundary diffusion has been indicated to dominate the transport mechanism [8]. Thus for isothermal deformation the simplified expression now appears as:

$$\qquad (2)$$

$$\dot{\varepsilon} = \frac{A_{II}\left(\sigma_e\right)^{\left(\frac{1}{M}\right)}}{d^a} + A_{III}\,\sigma^n$$

where A_{II} and A_{III} are constants, d is the average grain size, a is the grain size exponent, σ_e is the effective stress ($\sigma_e = \sigma - \sigma_o$), M is the maximum strain rate sensitivity, and n is the power law exponent. The application of this form of a constitutive relation to the SPD of Ti-6Al-4V has been previously discussed [6].

The constants in this constitutive relation must be quantified in order to represent the strain and strain-rate hardening behavior of Ti-6Al-4V at 900C.

Grain Growth

Stress/strain rate behavior of Ti-6Al-4V is well known to be strongly dependent on grain size [9]. It has been shown that grain coarsening occurs during deformation and causes flow hardening as well as changes in the m value [10]. To describe grain growth, both static grain growth (SGG) and deformation-enhanced grain growth (DEGG) effects were measured experimentally and represented by appropriate relations in order to properly incorporate this effect in superplastic flow equation (2).

Static Grain Growth

Static grain growth was determined by measuring grain size after isothermal exposure at 850, 875, 900, & 925C for times of 15 minutes to 24 hours. The samples were encapsulated under vacuum in quartz tubes for the thermal exposure. Average grain sizes were measured using the Abrams Circle method following ASTM E112 guidelines. Since our constitutive relation

5

contains stress and grain size as state variables, we use the following description of static grain growth rate:

$$\dot{d}_s = \frac{B}{d^P}$$ (3)

where d is the current grain size, P and B are constants for isothermal growth, and $\dot{d}_s$ is the static grain growth rate. The parameters P and B were determined by least squares fit to the experimental data.

Deformation Enhanced Grain Growth

As shown in Figure 1, superplastic deformation enhances the grain growth. There are several mechanistic models proposed to describe DEGG including grain switching [11], grain sliding/migration [11], grain cellular dislocation glide [12], grain cellular dislocation climb [13], and enhanced grain boundary mobility [14]. The DEGG models can be expressed as:

$$\dot{d}_d = \lambda \, d \, \dot{\varepsilon}$$ (4)

where $\lambda = \lambda(\dot{\varepsilon}, \varepsilon)$ in general. All but the latter two models require that λ be constant; however experimental results clearly show that λ is not constant (Figure 2).

We have chosen to use a model based on work by Clark and Alden [14] since it allows for a non-constant λ and can be adapted to microduplex microstructures [15]:

$$d^{(P+1)} = K_4 \, \varepsilon - K_4 \, \tau \, \dot{\varepsilon} \left(1 - \exp\left(- \frac{\varepsilon}{\dot{\varepsilon}\tau}\right) \right)$$ (5)

The $\left(1 - \exp\left(- \frac{\varepsilon}{\dot{\varepsilon}\tau}\right)\right)$ term indicates the generation and annihilation of 'excess' vacancies. Both K_4 and τ are constants which are determined by least square fit of the equation to the experimental grain growth data. When (5) is differentiated with respect to time (under constant strain rate conditions) an expression for DEGG is derived.

The resulting expression for total grain growth is dependent on strain and strain rate and is shown in equation (6):

$$\dot{d}_T = \dot{d}_s + \dot{d}_d = \frac{B}{d^P} + \frac{K_4 \, \tau \, \dot{\varepsilon}}{d^P} \left(1 - \exp\left(- \frac{\varepsilon}{\dot{\varepsilon}\tau}\right) \right)$$ (6)

Figure 3 shows our predictions of grain size with strain at 900C over a range of strain rates. This relation for total grain growth is used in the SPF constitutive relation (2).

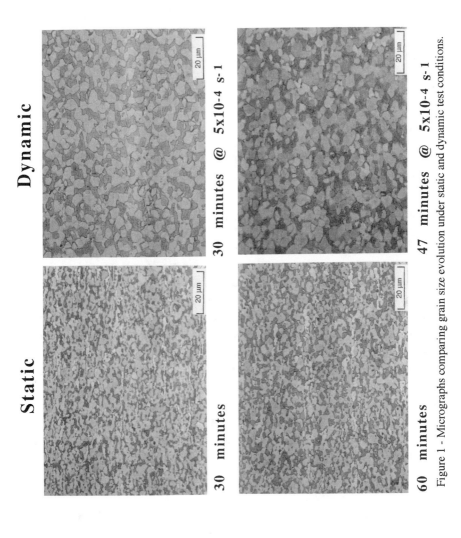

Dynamic

Static

30 minutes @ 5x10-4 s-1

30 minutes

47 minutes @ 5x10-4 s-1

60 minutes

Figure 1 - Micrographs comparing grain size evolution under static and dynamic test conditions.

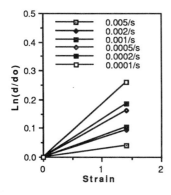 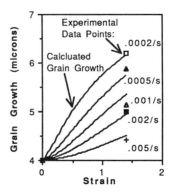

Figure 2 - Experimental data shown as
ln(d/do) vs. strain at 900C.

Figure 3 - Computed and experimental grain
size vs. strain at 900C.

Mechanical Parameters

The initial grain size, corresponding to the start of deformation, was measured after heating a
sample to test temperature, but before any strain was imposed. Experimental stress/strain and
stress/strain-rate data were used to determine the parameters M, σ_o, A_{II}, and A_{III}. The
threshold stress was measured in a stress relaxation test. Both M and the pre-constants were
determined from an experimental data corresponding to $\varepsilon=0$ (Figure 4). Each stress value on
this plot was determined from individual tensile tests conducted at various constant strain rates.
The experimental stress/strain data were corrected for elastic effects and machine compliance
effects in order to determine the strain corresponding to $\varepsilon=0$. This approach was used in order
to avoid any confusion with grain growth effects (as may happen with results from a step strain
rate test). The maximum strain rate sensitivity M, was directly measured from the curve in
Figure 4. The final set of parameters is listed in Table 1 below:

Table I Parameters for the constitutive relation at 900C

MECHANICAL PARAMETERS:		GRAIN GROWTH PARAMETERS:	
A_{II}	1.3×10^{-18}	B	0.0412 (microns/sec)
σ_o	0.25 (MPa)	P	3.9
a	3	K_4	939 (microns^P)
M	0.7	λ	1300 (seconds)
G	4.1×10^6 (N/cm^2)	d_o	4 (microns)
n	4.3		
A_{III}	6.0×10^9		

8

A comparison of the calculated stress/strain curves with the experimentally determined curves is shown in Figure 5. The constitutive relation with grain growth appears to provide a good representation of the flow behavior of the Ti-6Al-4V.

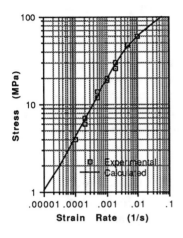
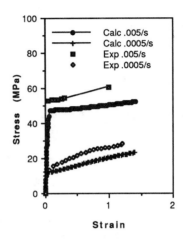

Figure 4 - Stress vs. strain rate showing experimental data and calculated values using the constitutive equation (900C).

Figure 5 - Stress vs. strain curves at constant strain rate comparing experimental data and calculated values (.005/s exp. data is corrected for necking; 900C).

Process Optimization

The constitutive relation was utilized to design a deformation rate profile intended to minimize the process time while preserving uniformity. The approach utilized here was to consider the conditions leading to the onset of instability locally, based on considerations of Hart and Nichols [16, 17]. This stability criterion is founded on the geometrical statement that if, in a homogenous section of a uniaxial test specimen, the change in growth rate of the cross-sectional area is positive, then it is unstable and the local neck will grow. This criterion is expressed in the mathematical form shown below:

$$m + \gamma = 1 \qquad\qquad\qquad (7)$$

$$\text{where } m \equiv \frac{\dot{\varepsilon}}{\sigma}\left(\frac{\partial\sigma}{\partial\dot{\varepsilon}}\right)_{\varepsilon} \qquad \text{and } \gamma \equiv \frac{1}{\sigma}\left(\frac{\partial\sigma}{\partial\varepsilon}\right)_{\dot{\varepsilon}}$$

where m is the strain rate sensitivity and γ is the strain hardening parameter. Both m and γ are functions of strain and strain rate for the Ti alloy, and can be computed from the constitutive relation (2). However, since the hardening is a result of grain growth only, an analogous set

9

of material parameters and stability criterion can be established which is based on grain size changes:

m' + γ' = 1

$$\text{where } m' \equiv \frac{\sigma}{\dot{\varepsilon}}\left(\frac{\partial\dot{\varepsilon}}{\partial\sigma}\right)_d \qquad \text{and } \gamma' \equiv \frac{\dot{d}}{\dot{\varepsilon}^2}\left(\frac{\partial\dot{\varepsilon}}{\partial d}\right)_\sigma \qquad (8)$$

where m' is the inverse strain rate sensitivity and γ' is the grain hardening parameter. In order to design a strain rate path for maximizing deformation rate while maintaining resistance to necking, a computer code was written. The program was designed to compute m and γ and iterate to solve for the largest average strain rate corresponding to the instability criterion. The computations were made for small strain increments (~0.02) and continued to a desired maximum strain (~1.5). All calculations were based on the experimental test specimen which had an initial gage length of 0.5 inches (shoulder-to-shoulder). The resulting strain rate path is shown in Figure 6. This data was used by a computer to control the crosshead of the test machine.

Experimental tests of the concept were conducted using an Instron (model TTL) tensile test machine modified with a computer controlled stepper motor. The specimen was wrapped in tantalum and contained in an Inconel 600 retort flowing with argon, and heated by a three-zone furnace to 900° +/-2C. A three hour heat up was used to stabilize the system, and a furnace cool was imposed on the specimen after testing. The specimen was subsequently reheated to 900C for five minutes and water quenched to re-establish the microstructure present at 900C. All test data were collected on the computer (usually one data point per second) with a concurrent load-displacement plot, and post-test data of time, load, displacement, stress, and strain.

Tensile tests were conducted to a strain of 1.2 (a strain considered to be representative of most superplastically formed parts). Figure 7 shows the experimental stress/strain data for the optimized deformation, compared with the prediction utilizing the constitutive relation (2). The representation appears valid.

In Figure 8 the optimized deformation path is shown, as well as various constant strain rate paths. The similarity of the failure envelope of the constant strain rate tests, to the shape of the optimum path supports the validity of using a criterion as done in this research. A comparison of time to desired strain is apparent on this plot. For example, following the optimum path to a strain of 1.2 required 950 seconds, whereas a constant strain rate of 0.0005/s required 2400 seconds. The strain rate of 0.0005/s or lower is approximately that used for many industrial forming processes.

10

Figure 9 shows a comparison of strains based on measured post-test areas along the gage of test specimens. A constant strain rate of 0.0005/s shows very uniform deformation, while a

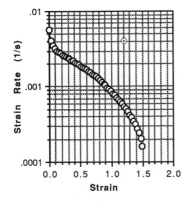
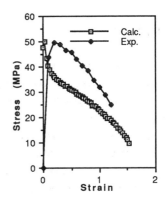

Figure 6 - Optimum strain rate path; computed using the constitutive equation and stability criterion at 900C.

Figure 7 - Stress vs. strain curve comparing the experimental data from the optimized deformation path, and the calculated values.

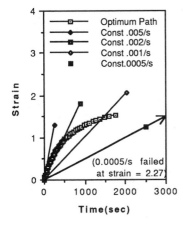
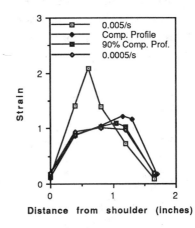

Figure 8 - Strain vs. time profiles for various constant strain rates (ending at failure) and the optimum (designed) deformation path corresponding to Figure 6 (at 900C).

Figure 9 - Measured local strain along gage length of the test specimens after deformation to an average longitudinal strain of 1.2 (230% elongation, 900C).

constant strain rate of 0.005/s displays local necking. The optimum profile showed some necking, so a run was made with a profile at 90% of calculated strain rate (more conservative) and the specimen approached the uniformity of the 0.0005/s test.

Finite Element Modeling

The previous work was based on the assumption at uniformity along the specimen (shoulder to shoulder). However, strain varies along this gage length during deformation due to variations in stress state (i.e. plane strain at the shoulder to plane stress at the midsection). Material from the grip area in the specimen also flows into the gage section during the test. To account for this inhomogenous behavior, finite element modeling (FEM) was utilized. This technique utilized an element of 0.01" rather than the total gage length of 0.5" to design the tensile test. The FEM software has a front end which iterates the crosshead velocity to meet the stability criteria for one or more assigned elements. The predicted local strain rate meeting the stability criteria is entered as a fifth order polynomial. The FEM iterates using the secant method with explicit integration (Newton forward). This results in a stable, fast converging program.

Figure 10 shows the crosshead velocity vs. time profile for the FEM compared to that generated with equation (7). The FEM path is generally more conservative and compares closely with the (arbitrary) 90% path previously tested. A test to a strain of 1.2 required 1170 seconds. Figure 11 shows the specimen uniformity in terms of area strain along the gage length (as done previously) for a specimen tested following the optimum velocity profile established with the FEM procedure.

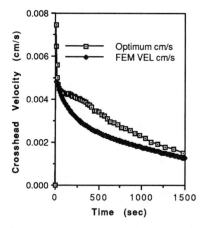

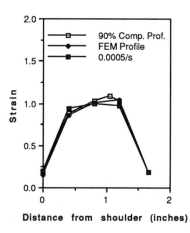

Figure 10 - Crosshead velocity vs. time control profiles are compared between the optimized and the FEM designed tests.

Figure 11 - Local strain along gage length after deformation to an average longitudinal strain of 1.2 (230% elongation, 900C).

12

Summary and Conclusions

The concept of designing a deformation rate path for increasing superplastic forming rates appears to have merit based on the study conducted. By utilizing these designed variable strain rate paths, tensile samples of Ti-6Al-4V were deformed to an average strain of 1.2 in 950 seconds as compared to 2400 seconds for the constant strain rate of 0.0005/s. The resulting thinning in both tests is virtually identical. By invoking an instability criterion, it appears that a maximum strain rate can be imposed which will not cause a damaging extent of necking. While other research has shown similar promising results [18], this study appears to be the first demonstrating process design using a constitutive relation, rather than experimentally measured strain and strain-rate hardening. This can be significant since these mechanical properties may be path dependent, creating uncertainties resulting from the experimental route to determine the parameters. That is, the grain size and resulting values of m and γ at a given strain will depend on the deformation history used to develop the strain. This approach should also be amenable to process design for part forming.

The following conclusions are drawn from this study:

1. A constitutive relation was developed which represented the superplastic behavior of Ti-6Al-4V at 900C. Flow stress, strain hardening and strain rate sensitivity were correlated for verification.

2. Grain growth kinetics can be represented by appropriate mechanistic models. Both static and dynamic effects must be considered to accurately represent the development of the microstructure.

3. The constitutive relation, along with a stability criterion, can be used to generate a strain rate vs. strain, 'optimized' deformation path for minimizing deformation time. This path appears to meet the intent of establishing a relatively uniform deformation locally. The use of FEM in conjunction with the stability criterion appears to be the preferred analytical method.

4. Designing strain rate profiles for superplastic deformation can result in significant deformation time savings.

Acknowledgments

The authors acknowledge and appreciate support for various aspects of this research provided by the National Science Foundation (Grant #DDM-9914852, Dr. Bruce M. Kramer, Cognizant Program Official), British Aerospace Co., and General Electric Co. The principal author would like to thank the graduate school of Washington State University, and the Department of Mechanical and Materials Engineering for their financial support.

References

1. J. Wadsworth, T. G. Nieh, and O. D. Sherby, "Future Directions for Superplasticity", International Conference on Superplasticity in Advanced Materials, Ed. S. Hori, M. Tokizane and N. Furushiro. (Osaka, Japan: Japan Society for Research on Superplasticity, 1991) 13-22.

2. C. H. Hamilton et al., "Microstructural Coarsening and its Effect on Localization of Flow in Superplastic Deformation", International Conference on Superplasticity in Advanced Materials, ICSMA, Ed. Norio Furushiro, Shigenori Hori and Masahara Tokzane. (Osaka, Japan: Japan Society for Research on Superplasticity, 1991) 127-132.

3. B. A. Ash, and C. H. Hamilton, "Strain and Strain Rate Hardening Characteristics of a Superplastic Al-Zr-Cu-Zr Alloy," Scripta Metall. 22 (1988): 277-282.

4. John Pilling, and Norman Ridley, Superplasticity in Crystalline Solids (Southampton: The Institute of Metals, 1989) 214.

5. M. F. Ashby, and R. A. Verrall, "Diffusion-Accommodated Flow and Superplasticity," Acta Metall. 21 (1973): 149-163.

6. C. H. Hamilton et al., "Dynamic Grain Coarsening and its Effect on Flow Localization in Superplastic Deformation", The Second International SAMPE Symposium, (Chiba, Japan: Society for the Advancement of Material and Process Engineering, 1991) 272-279.

7. B. P. Kashyap, and A. K. Mukherjee, "On the Models for Superplastic Deformation", Superplasticity, Ed. B. Baudelet and M. Suery. (Grenoble, France: Centre National de la Recherche Scientifique, Paris, 1985) 4.1-4.27.

8. H. J. Frost, and M. F. Ashby, Deformation - Mechanism Maps (Pergamon Press, 1982)

9. D. Lee, and W. A. Backofen, "Superplasticity in Some Titanium and Zirconium Alloys," Trans. TMS-AIME 239.July (1967): 1034-1040.

10. A. K. Ghosh, and C. H. Hamilton, "Influences of Material Parameters and Microstructure on Superplastic Forming," Metall. Trans. 13A (1982): 733-741.

11. D. S. Wilkinson, and C. H. Caceres, "On the Mechanism of Strain-Enhanced Grain Growth During Superplastic Deformation," Acta Metall. 32.9 (1984): 1335-1345.

12. E. Sato, K. Kuribayashi, and R. Horiuchi, "A Mechanism of Superplastic Deformation and Deformation Induced Grain Growth Based on Grain Switching", Superplasticity in Metals, Ceramics and Intermetallics, MRS, 1990) 196: 27-32.

13. D. J. Sherwood, and C. H. Hamilton, "A Mechanism for Deformation-Enhanced Grain Growth in Single Phase Materials," Scripta Metallurgica 25 (1991): 2873-2878.

14. M. A. Clark, and T. H. Alden, "Deformation Enhanced Grain Growth in a Superplastic Sn-1%Bi Alloy," Acta Metall. 21 (1973): 1195-1206.

15. O. N. Senkov, and M. M. Myshlyaev, "Grain Growth in a Superplastic Zn-22%Al Alloy," Acta Metall. 34 (1986): 97-106.

16. E. W. Hart, "Theory of the Tensile Test," Acta Metallurgica 15 (1967): 351-355.

17. F. A. Nichols, "Plastic Instabilities and uniaxial Tensile Ductilities," Acta Metallurgica 28 (1980): 663-673.

18. B. Ren, C. H. Hamilton, and B. A. Ash, "An Approach to Rapid SPF of an Al-Li-Cu-Zr Alloy", Fifth International Aluminum-Lithium Conference, Ed. Jr. E. A. Starke and T. H. Sanders, Jr. (Williamsburg, VA: The Metallurgical Society, 1989)

INSTABILITY ANALYSIS FOR DYNAMICALLY RECRYSTALLIZED

SUPERPLASTIC ALUMINUM-LITHIUM ALLOYS

M.N. Srinivasan and R.E. Goforth

Department of Mechanical Engineering
Texas A&M University
College Station, Texas 77843-3123

ABSTRACT

The limit of deformation in superplastic forming is often governed by tensile instability since stretching and blow forming are common methods of superplastic forming. In fully recrystallized superplastic alloys, strain rate hardening is the primary source of strengthening against tensile instability, while in dynamically recrystallized alloys, both strain hardening and strain rate hardening contribute to the strengthening against necking. Several aluminum-lithium alloys have been developed for superplastic forming and three important members of this group are ALCOA 2090-OE16, Weldalite 049 and Alcan 8090 SP. In this paper the instability behavior of these three alloys have been examined with respect to changes in strain rate, gage thickness and testing temperature. Attempts have also been made to relate the instability to the tensile elongation and to offer qualitative explanations for the differences in the instability by considering the different factors affecting the relationship between the Instability Parameter and the tensile strain.

Advances in Superplasticity and Superplastic Forming
Edited by N. Chandra, H. Garmestani, R.E. Goforth
The Minerals, Metals & Materials Society, 1993

INTRODUCTION

Investigations on the superplastic behavior of three different dynamically recrystallized aluminum-lithium alloys have been carried out in the Deformation Processing Laboratory of Texas A&M University over the last four years [1-9]. The main objective of this program was to develop constitutive equations which best describe the relationship between stress, strain, strain-rate and temperature for each alloy. These equations are necessary for the prediction of optimum pressure-time cycle and thickness changes using a finite element model [10,11]. In addition, each tested sample was subjected to detailed microstructural analysis using light, scanning electron and transmission electron microscopes. Furthermore, each stress-strain curve was analyzed to determine the strain hardening index and the strain rate sensitivity factor in order to determine the Instability Parameter. In a tensile test specimen, instability is initiated at a strain where the instability parameter exceeds unity, but whether the specimen fails at this strain will depend upon the extent of counteraction by strengthening factors. The rationale for initiation of instability is briefly discussed below.

In a metallic material, dislocations are responsible for strain hardening once the material is stressed beyond the yield point. If the material can be superplastically processed, the plastic deformation is also governed by the strain rate hardening. Micrograin superplasticity has been achieved in both statically recrystallized and dynamically recrystallized alloys. In the former, the alloy is fully recrystallized to a fine equiaxed grain size prior to superplastic deformation. In the latter, the alloy is heavily warm worked and left unrecrystallized, so that strain-induced recrystallization occurs during the superplastic deformation. It is generally accepted that the dislocation density in dynamically recrystallized alloys is much greater than in the fully recrystallized alloys. It is therefore to be surmised that the dynamically recrystallized alloys are affected by both strain hardening and strain rate hardening, while the statically recrystallized alloys are predominantly affected by strain rate hardening, because of the relatively low dislocation density. The tensile instability in dynamically recrystallized alloys is therefore to be determined in terms of both strain hardening and strain rate hardening effects. The following procedure may be adopted for this purpose [12-14].

Based on the nomenclature provided at the end of this paper, instability is initiated in a tensile specimen when

$$\frac{dP}{dL} = 0 = A \left\{ \left(\frac{\delta \sigma}{\delta \epsilon} \right)_{\dot{\epsilon}} \frac{d\epsilon}{dL} + \left(\frac{\delta \sigma}{\delta \dot{\epsilon}} \right)_{\epsilon} \frac{d\dot{\epsilon}}{dL} \right\} + \sigma \frac{dA}{dL} \qquad (1)$$

It can be shown that in equation (1)

$$\frac{d\epsilon}{dL} = -\frac{1}{A} \frac{dA}{dL} \qquad (2)$$

and

$$\frac{d\dot{\epsilon}}{dL} = -\frac{1}{A} d \frac{\dot{A}}{dL} + \frac{\dot{A}}{A^2} \frac{dA}{dL} \qquad (3)$$

In equation (1), $\delta\sigma/\delta\epsilon$ represents the strain hardening index, which can be obtained from the flow stress - strain curve for a given strain rate, temperature and gage thickness. However, it is more convenient to use the dimensionless strain hardening coefficient which is determined as

18

$$\gamma = \frac{1}{\sigma} \frac{\delta\sigma}{\delta\epsilon} \qquad\qquad (4)$$

In general, there is a sigmoidal relationship between the logarithm of the flow stress and the logarithm of the strain rate and the value

$$m = \left(\frac{\delta\ln\sigma}{\delta\ln\dot{\epsilon}}\right)_{\epsilon} = \frac{\dot{\epsilon}}{\sigma}\left(\frac{\delta\sigma}{\delta\epsilon}\right)_{\dot{\epsilon}} \qquad\qquad (5)$$

is known as the strain rate sensitivity factor.

Substitution of equations (2) to (5) in (1) and rearrangement of the terms results in

$$\frac{\left[\dfrac{d(\ln \dot{A})}{dL}\right]}{\left[\dfrac{d(\ln A)}{dL}\right]} = \frac{m+\gamma-1}{m} \qquad\qquad (6)$$

Instability (necking) in a tensile test is initiated when at any length L in the gage, the rate of change of area with time is smaller than the change in the actual area at this period. It is clear from equation (6) that this condition develops when $\left[\dfrac{1-m-\gamma}{m}\right]$ is greater than zero. The term in the parenthesis has been defined as the 'Instability Parameter' by Hamilton, et al [12], who used this parameter to study the instability of a dynamically recrystallized Al-Li-Zr superplastic alloy. However, their study was on only one gage thickness and the data was provided only at one temperature [12]. As noted earlier, the present study includes three different dynamically recrystallized alloys tested at different combinations of strain rate, gage thickness and temperature, which enables instability to be analyzed in greater detail.

EXPERIMENTAL

Three dynamically recrystallized superplastic aluminum-lithium alloys were studied in the present work. They were ALCOA 2090-OE16 (2.2% Li, 2.7% Cu, 0.12% Zr, bal Al), Weldalite 049 (1.3% Li, 4.75% Cu, 0.4% Ag, 0.4% Mg, 0.14% Zr, 0.03% Ti, bal Al) and Alcan 8090-SP (2.5% Li, 1.2% Cu, 0.7% Mg, 0.12% Zr, bal Al). 2090-OE16 samples were made from sheets of 0.063 in (1.60 mm), 0.090 in (2.29 mm) and 0.125 in (3.18 mm) thickness, while 8090-SP samples were made from sheets of 0.063 in (1.6 mm), 0.090 in (2.29 mm) and 0.118 in (3.0 mm) thickness. Weldalite 049 sheet was available only in 0.090 in (2.29 mm) thickness. The gage thickness of each tensile sample was 0.25 in (6.35 mm) and the gage width was 0.185 in (4.7 mm). All the samples were tested in the rolled (longitudinal) direction of the sheet.

The different combinations of variables studied in the present work are listed in Table 1.

Table 1 - Combination of Variables

Code	Alloy	Gage Thickness, in	Strain Rate, in/in/sec	Test Temp, C
A1	2090-OE16	0.090	0.0004	518
A2	2090-OE16	0.090	0.0008	518
A3	2090-OE16	0.090	0.0010	518
B1	2090-OE16	0.125	0.0004	518
B2	2090-OE16	0.125	0.0008	518
B3	2090-OE16	0.125	0.0010	518
C1	2090-OE16	0.063	0.0002	510
C2	2090-OE16	0.063	0.0005	510
C3	2090-OE16	0.063	0.0008	510
C4	2090-OE16	0.063	0.0016	510
D1	2090-OE16	0.090	0.0002	510
D2	2090-OE16	0.090	0.0005	510
D3	2090-OE16	0.090	0.0008	510
D4	2090-OE16	0.090	0.0016	510
E1	2090-OE16	0.125	0.0002	510
E2	2090-OE16	0.125	0.0004	510
E3	2090-OE16	0.125	0.0008	510
E4	2090-OE16	0.125	0.0016	510
F1	8090-SP	0.063	0.0004	518
F2	8090-SP	0.063	0.0008	518
F3	8090-SP	0.063	0.0010	518
G1	8090-SP	0.090	0.0004	518
G2	8090-SP	0.090	0.0008	518
G3	8090-SP	0.090	0.0010	518
H1	8090 SP	0.118	0.0004	518
H2	8090 SP	0.118	0.0008	518
H3	8090 SP	0.118	0.0010	518
I1	Weldalite 049	0.090	0.0002	470
I2	Weldalite 049	0.090	0.0004	470
I3	Weldalite 049	0.090	0.0006	470
I4	Weldalite 049	0.090	0.0008	470
J1	Weldalite 049	0.090	0.0002	490
J2	Weldalite 049	0.090	0.0004	490
J3	Weldalite 049	0.090	0.0006	490
K1	Weldalite 049	0.090	0.0002	510
K2	Weldalite 049	0.090	0.0004	510
K3	Weldalite 049	0.090	0.0006	510

Code	Alloy	Gage Thickness, in	Strain rate, in/in/sec	Test Temp, C
K4	Weldalite 049	0.090	0.0008	510
L1	Weldalite 049	0.090	0.0002	530
L2	Weldalite 049	0.090	0.0004	530
L3	Weldalite 049	0.090	0.0006	530
L4	Weldalite 049	0.090	0.0008	530

A schematic diagram of the superplastic tensile testing set-up is shown in Fig. 1 and in Fig. 2 is shown the view of a tensile specimen.

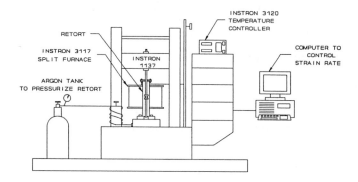

Figure 1 - Experimental Setup

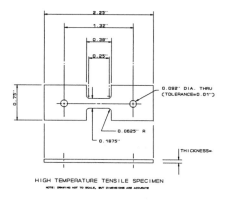

Figure 2 - Specimen Dimensions

The set-up consists of an INSTRON 1130 Universal Testing Machine controlled by a PC's Limited 286 Computer with a software specially developed for constant strain rate testing. The sample is housed in a stainless steel retort rated for 1000 psi and 800 C. The retort is water-cooled and made leak-proof so as to contain argon circulated under pressure during testing. The purpose of argon is to provide the necessary hydrostatic (back) pressure to minimize the effect of cavitation during deformation. In the present work all tests were conducted under a constant argon pressure of 400 psi (2.76 MPa). The retort is surrounded by a three-zone split furnace (INSTRON 3117) which can maintain the specimen temperature within ±3 C of the set temperature, with the aid of an INSTRON 3120-001 self-adaptive temperature controller. Hard copies of the stress-strain curves generated from the data stored in the computer were used to determine the strain to failure and also analyzed to determine the relationship between the flow stress and the strain rate at different strain levels. The dimensionless strain hardening index, γ, was determined by dividing the average slope of the stress - strain curve at different strains, by the flow stress at the particular strain. The strain rate sensitivity parameter, m, was determined as the average slope of the logarithmic flow stress versus the logarithmic strain rate curve at a given strain level. The instability parameter was determined from these values as described earlier.

RESULTS AND DISCUSSION

A. Instability

In Fig. 3 and Fig. 4 are shown typical variations of γ and m with strain. The instability parameter calculated from the values of γ and m at each strain is turn plotted against the strain in Fig. 5.

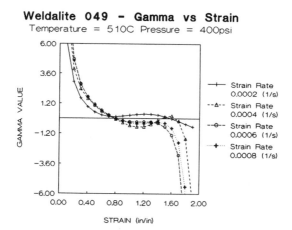

Figure 3 - Variation of γ with strain

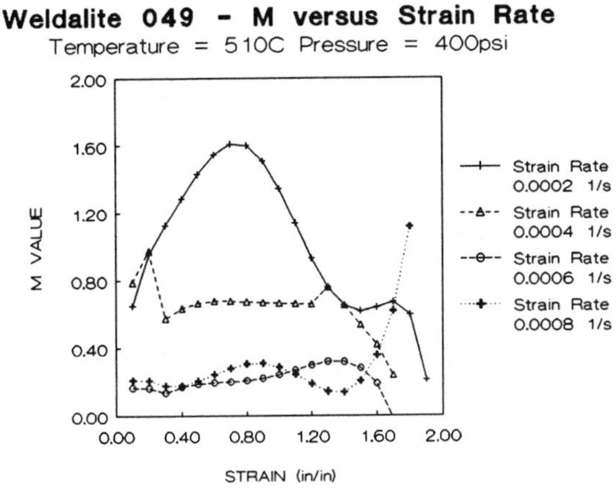
Weldalite 049 - M versus Strain Rate
Temperature = 510C Pressure = 400psi
M VALUE
2.00
1.60
1.20
0.80
0.40
0.00
0.00
0.40
0.80
1.20
1.60
2.00
STRAIN (in/in)
Strain Rate 0.0002 1/s
Strain Rate 0.0004 1/s
Strain Rate 0.0006 1/s
Strain Rate 0.0008 1/s

Figure 4 - Variation of m with strain

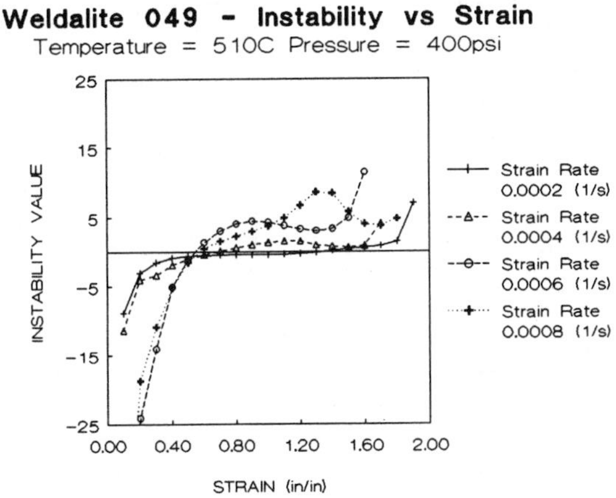
Weldalite 049 - Instability vs Strain
Temperature = 510C Pressure = 400psi
INSTABILITY VALUE
25
15
5
-5
-15
-25
0.00
0.40
0.80
1.20
1.60
2.00
STRAIN (in/in)
Strain Rate 0.0002 (1/s)
Strain Rate 0.0004 (1/s)
Strain Rate 0.0006 (1/s)
Strain Rate 0.0008 (1/s)

Figure 5 - Variation of I with strain

The shape of the Instability Parameter - Strain curve differs amongst the different samples in the following respects: the strain at initial instability, the shape of the curve after the onset of initial instability - indicating the dominance of hardening or softening or balance between the two, the strain at final instability and the maximum value of the Instability Parameter during deformation. In order to make effective comparisons amongst the different samples in respect to the instability behavior therefore, all the above data need to be extracted from the Instability Parameter - Strain curve for each sample. This data is recorded in Table 2 and Table 3. For convenience, the elongation of each sample is also listed in Table 2. The equivalent true strain is listed in Table 4, for comparison with the strain at final instability.

Table 2 - Strain at Instability

Code	Strain At Initial Instability (A)	Strain At Final Instability (B)	Elongation, %	Hardening Or Softening After (A)
A1	0.64	1.94	1373	Balance
A2	0.78	1.83	1011	Balance
A3	0.78	1.61	1098	Balance
B1	0.69	1.84	883	Balance
B2	0.69	1.93	589	Balance
B3	0.59	1.88	555	Balance
C1	0.35	2.07	1454	Balance
C2	0.37	2.05	1071	Balance
C3	0.37	1.93	1094	Balance
C4	0.38	1.74	725	Balance
D1	0.45	2.47	1690	Hardening
D2	0.49	1.87	1600	Hardening
D3	0.55	2.00	1047	Balance
D4	0.55	1.74	839	Balance
E1	0.53	2.40	1359	Balance
E2	0.54	2.13	1094	Balance
E3	0.80	1.33	740	Softening
E4	0.80	1.02	400	Softening
F1	0.66	1.94	750	Balance
F2	1.07	2.18	785	Balance
F3	0.83	1.83	523	Softening
G1	0.74	2.36	963	Hardening
G2	0.91	1.50	591	Balance
G3	1.07	1.32	585	Softening
H1	0.66	1.87	882	Balance
H2	0.61	1.79	818	Balance
H3	0.72	1.67	770	Balance
I1	0.68	2.30	898	Hardening
I2	0.68	1.68	637	Balance
I3	0.68	1.68	579	Balance
I4	0.68	1.14	477	Softening
J1	0.61	2.62	1275	Balance
J2	0.48	1.56	496	Balance

Code	Strain At Initial Instability (A)	Strain At Final Instability (B)	Elongation, %	Hardening Or Softening After (A)
J3	0.61	1.47	463	Balance
K1	1.39	1.77	1080	Hardening
K2	0.69	1.56	617	Balance
K3	0.53	1.45	542	Balance
K4	0.53	1.79	564	Balance
L1	1.09	1.60	740	Balance
L2	0.63	1.41	496	Balance
L3	0.55	1.26	403	Balance
L4	0.51	0.57	326	Softening

Table 3 - Maximum value of the Instability Parameter

Code	I_{max}	Code	I_{max}	Code	I_{max}
A1	0.78	E1	4.29	I2	4.54
A2	0.52	E2	3.14	I3	7.72
A3	0.00	E3	6.57	I4	25.0
B1	2.50	E4	20.0	J1	0.83
B2	2.36	F1	1.11	J2	1.38
B3	2.36	F2	1.53	J3	50.0
C1	9.23	F3	0.56	K1	6.82
C2	1.54	G1	0.56	K2	4.09
C3	0.00	G2	10.0	K3	11.4
C4	0.00	G3	10.0	K4	9.09
D1	2.82	H1	2.64	L1	3.64
D2	2.11	H2	8.63	L2	1.36
D3	1.06	H3	3.75	L3	4.54
D4	0.00	I1	0.00	L4	12.7

Table 4 - Strain at Fracture

Code	ϵ_f	Code	ϵ_f	Code	ϵ_f
A1	2.69	E1	2.68	I2	2.00
A2	2.41	E2	2.48	I3	1.92
A3	2.48	E3	2.13	I4	1.75
B1	2.29	E4	1.61	J1	2.62
B2	1.93	F1	2.14	J2	1.79
B3	1.88	F2	2.18	J3	1.73
C1	2.74	F3	1.83	K1	2.47
C2	2.46	G1	2.36	K2	1.97

Code	ϵ_f	Code	ϵ_f	Code	ϵ_f
C3	2.48	G2	1.93	K3	1.86
C4	2.11	G3	1.92	K4	1.89
D1	2.88	H1	2.28	L1	2.13
D2	2.83	H2	2.22	L2	1.79
D3	2.44	H3	2.16	L3	1.61
D4	2.24	I1	2.30	L4	1.45

In order for a specimen to attain a large value of the strain at fracture (corresponding to the percentage elongation), the following conditions may be considered to be favorable.

1. Large strain at initial instability
2. Dominance of hardening process after initial instability, or at least, balance between softening and hardening events after initial instability
3. Low maximum instability parameter after initial instability
4. Large difference between the strains at initial and final instability
5. Small difference between the strain at final instability and the strain at fracture.

It is noted from the analysis of these values in Table 2, Table 3 and Table 4 that all the above noted conditions are not met in the specimen that showed the best elongation in this series (D1). Though the favorable conditions are that there is dominance of hardening after initial instability, the difference between the strains at initial and final instability is large (2.02) and the difference between the strain at final instability and the strain at fracture is small (0.41), the unfavorable conditions are that the maximum value of the instability parameter is fairly large (2.82) and the strain at initial instability is low (0.45). A plausible explanation for this observation is that the negative effects are overshadowed by the strong showing in the other three conditions in this particular specimen. In other words, the five factors listed above may have different weightages depending upon the sample.

It is seen in Table 2, Table 3 and Table 4 that there is no general trend in the variation of the strain at initial instability, the strain at final instability, the dominance of hardening or softening after initial instability or the strain at fracture with respect to the superplastic variables. The reason for the lack of any regular trend may be explained as follows. It has been observed by the studies Texas A&M University (2,5,8) that the transmission electron microstructures of all the three alloys studied show the presence of hexagonal subcells which undergo dynamic (continuous) changes in dimensions as a result of dynamic recovery. Such subcells have been noted in other alloys and have been attributed to stacking fault (15). Fig. 6 is a typical transmission electron microscope photograph showing the presence of subcells.

Figure 6 - TEM photograph of 2090-OE16
showing subcells   x 250,000

The fact that dynamic recovery is associated with changes in the dimensions of the subcells was confirmed by the present authors through corresponding selected area diffraction (SAD) photographs which indicated different patterns associated with different energy levels with progressive strain. Since the size of the subcells varies in a random fashion with the strain, it is implied that the interaction between the hardening and softening processes is also random and the relative weightages of the instability factors noted above may vary from one specimen to another, resulting in a lack of any general trend with regard to the superplastic variables.

It thus seems to be necessary to relate each of the factors noted above to corresponding changes in the microstructure to assess the relative weightage of each factor. It may then be possible to precisely explain the variations of the magnitudes of the factors in the different samples examined. The general statement that can be made from the analysis of the results in Table 2, Table 3 and Table 4 is that the instability would be high when several factors are adverse. For example, the specimen L4 that shows the lowest elongation in the series, has a maximum instability parameter value of 12.7, a small difference between the strains at initial and final instability (0.06), softening after the initial instability and a large difference between the strain at final instability and the strain at fracture (0.88).

Comparison of the elongations and the instability behavior at similar strain rates and temperatures indicates that 2090-OE16 has the best stability followed by 8090 SP and Weldalite 049. It has been observed in the studies at Texas A&M University (2,5,8) that the subcells grow to larger dimensions in 2090-OE16 before another cycle of dynamic recovery advances, as compared to 8090 SP and Weldalite 049, which seem to have comparable maximum cell sizes. Dynamic recovery appears to be somewhat faster in 8090 SP than in Weldalite 049. Further studies, however are needed before more quantitative statements can be made on the relative behavior of the alloys.

It may also be noted in this context that the Instability Parameter - Strain curve obtained by Hamilton et al [12] show that I=0 is maintained for a fairly wide strain range at all the strain rates considered by them. Though a few samples in the present work approach this trend, most samples show significantly higher than zero I_{max} values as seen in Table 4. This type of behavior has also been observed by Higashi, et al [16] in P/M 7475 aluminum alloy.

It is interesting to note that the initial instability and final instability have been attributed to diffuse necking and localized necking respectively [14]. In the present series it is difficult to determine if the final instability is caused by extended diffuse necking or localized necking, owing to the complex interaction between the hardening and softening processes.

B. Dual Strain Rate Application

It was first demonstrated by Hamilton, et al [12], that it is possible to strain a dynamically recrystallized superplastic alloy to a value less than or equal to the strain at initial instability at a relatively high strain rate and then lower the strain rate to the value that gives the highest elongation in the constant strain rate test, without sacrificing ductility. This procedure has the obvious advantage of reducing the test (or production) time.

In Table 5 are shown the elongations of dual strain rate tested 2090-OE16 and 8090 SP specimens. The initial strain rate was 0.002/s up to a strain of 0.5 in each case. At this stage the strain rate was decreased to 0.0002/s and maintained at this value until failure of the specimen.

Table 5 - Dual Strain Rate Elongations

Code	Alloy	Temperature C	Thickness in	Elongation %
DS1	2090-OE16	518	0.063	1092
DS2	2090-OE16	518	0.090	1461
DS3	2090-OE16	518	0.125	1077
DS4	8090 SP	518	0.063	1188
DS5	8090 SP	518	0.090	983
DS6	8090 SP	518	0.118	1222

It is clear from Table 5 that the elongations obtained after dual strain rate testing are good. If these elongations are compared with the constant strain rate elongations shown in Table 2 and allowances are made for differences in temperature or strain rate, it is indicated that there is practically no sacrifice in the ductility when the dual strain rate procedure is adopted, both in the case of 2090 OE16 and 8090 SP alloys. A recent study has indicated that a similar behavior is found in Weldalite 049 as well [5].

CONCLUSIONS

1. The instability parameter variation with strain during superplastic tests of dynamically recrystallized aluminum-lithium alloys shows considerable differences depending upon the alloy, strain rate, temperature and gage thickness.
2. ALCOA 2090-OE16 seems to display overall superior stability behavior when compared with Alcan 8090 SP or Reynolds Weldalite 049.
3. Some factors that promote stability against necking are, large strain at initial instability, large difference between the strains at initial and final instability, low maximum value of the instability parameter after the onset of initial instability, small difference between the strain at final instability and the strain at fracture. All these factors are affected by the microstructural dynamics and therefore, there is a clear need for extensive studies relating the microstructure to the above factors.

4. By using an initially high strain rate up to or below the strain at initial instability and then lowering it to a low (optimum) rate up to the point of failure, it is possible to obtain similar order of ductility as with the low strain rate alone. This procedure can therefore lead to significant reduction in production time without sacrificing ductility in these dynamically recrystallized alloys.

REFERENCES

1. L. Douskos, "Superplastic Behavior of Al-Li 2090 OE16 (ALCOA)," M.S. Thesis, Texas A&M University, May 1991
2. R. Balasubramanian, "Microstructural Characterization of Superplastic Aluminum-Lithium Alloys," M.S. Thesis, Texas A&M University, Aug 1990
3. R. Khazi-Syed, "Superplastic Forming Characteristics of 2090 (OE16) and 8090 (SP) Al-Li Alloys," M.S. Thesis, Texas A&M University, December 1991
4. J.R. Seldenrust, "Superplastic and Microstructural Characterization of Weldalite 049," M.S. Thesis, Texas A&M University, December 1991
5. V. Kalahasti, "Microstructural Characterization of Superplasticity of Weldalite 049," M.S. Thesis, Texas A&M University, Aug 1992
6. R.E. Goforth, M.N. Srinivasan, N. Chandra and L. Douskos, "Superplastic Flow Characteristics and Microstructural Analysis of Aluminum-Lithium Alloy 2090-OE16," in Superplasticity in Aerospace II, T.R. McNelley and H.C. Heikkennen (eds), TMS, Warrendale, PA, 285-302 (1990)
7. R.E. Goforth, M.N. Srinivasan and N. Chandra, "A Comparative Study of the Superplastic Flow Characteristics and Microstructural Analysis of Two Dynamically Recrystallizing Aluminum-Lithium Alloys," in Superplasticity in Advanced Materials, S. Hori, M. Tokizane and N. Furushiro (eds), The Japan Society for Research on Superplasticity, 145-150, (1991)
8. M.N. Srinivasan, R.E. Goforth and R. Balasubramanian, "Microstructural Evaluation of a Dynamically Recrystallized Aluminum-Lithium Alloy," Materials Characterization, Vol 27, No.3, (1992)
9. R.E. Goforth and M.N. Srinivasan, "Testing the Superplastic Flow Characteristics of Advanced Aluminum Alloys," Journal of Testing and Evaluation, ASTM, Philadelphia, PA, Vol 21, No. 1, 36-43, (1993)
10. N. Chandra, K. Chandy and S. Rama, "Computational Model for Superplastic Pans of Complex Geometry with Friction," op cit, Ref. 6, 67-86
11. S. Ramalingam, "Finite Element Analysis and Design of 3-Dimensional Superplastic Sheet Forming Processes" Ph.D. Thesis, Texas A&M University, 1992
12. C.H. Hamilton, B.A. Ash, D. Sherwood and H.C. Heikkennen, "Effect of Microstructural Dynamics on Superplasticity in Al Alloys," in Superplasticity in Aerospace, H.C. Heikkennen and T.R. McNelley (eds), TMS, Warrendale, PA, 29-50 (1988)
13. E.W. Hart, Acta Metallurgica, Vol 15, 351 (1967)
14. G.E. Dieter, "Mechanical Metallurgy", Chapter 8, McGraw-Hill, New York, 3rd edition, (1986)
15. L.M. Howe, M. Rainville and E.M. Schulson, "Transmission Electron Microscopy Investigations of Ordered Zr_3Al," Journal of Nuclear Materials, Vol 50, 139-154, (1974)
16. K. Higashi, H. Imamura, T. Ito and S. Tanimura, "The Instability of Superplastic Deformation in P/M and I/M 7475 Aluminum Alloys," op cit, Ref. 6, 223-234.

NOMENCLATURE

P - Tensile load

L - Length of the specimen along the gage

A - Cross sectional area of the specimen at the neck

$\dot{A}$ - Rate of change of cross sectional area at the neck

σ - Tensile stress, MPa

ϵ - Tensile strain, in/in

$\dot{\epsilon}$ - Strain rate, in/in/s

t - Time, s

γ - Dimensionless strain hardening index

m - Strain rate sensitivity factor

I - Instability Parameter

SUPERPLASTIC FLOW CHARACTERISTICS OF DYNAMICALLY

RECRYSTALLIZING ALUMINUM-LITHIUM ALLOYS

R.E. Goforth and M.N. Srinivasan

Department of Mechanical Engineering
Texas A&M University
College Station, Texas 77843-3123

ABSTRACT

Currently there is not a satisfactory explanation of the mechanisms involved in micrograin superplasticity. Several theories have been proposed and certain researchers favor particular theories, however, the complexity of the process and often contradictory nature of reported results, prevents the universal acceptance of any one theory. Experimental results for two aluminum-lithium alloys are compared to superplastic theories and models proposed by Ashby-Verrall, Ball-Hutchinson, and Mukherjee.

Advances in Superplasticity and Superplastic Forming
Edited by N. Chandra, H. Garmestani, R.E. Goforth
The Minerals, Metals & Materials Society, 1993

INTRODUCTION

There are basically two types of superplasticity: microstructural and environmental. Microstructural superplasticity, often called micrograin or fine grained superplasticity is the most common type. It is observed in materials which exhibit a fine grain size when deformed at a controlled strain rate (usually in the 10^{-4} to 10^{-3} $\sec^{-1}$ range) and at a temperature greater the $0.5T_m$ (T_m is the absolute melting temperature). This method of deforming a material is easily implemented into a commercial environment which makes micrograin superplasticity a method of industrial importance. Environmental or transformational superplasticity is observed in some materials when the temperature is cycled simultaneously as the load is applied. The complexity of producing a commercial process with controlled simultaneous cycling of temperature and load makes environmental superplasticity unattractive from a commercial standpoint.

Micrograin superplastic alloys can be divided into two types: pseudo single phase and microduplex.[1] Pseudo single phase alloys are processed so as to have a fine grain size along with fine dispersoids distributed throughout the matrix. These dispersoids prevent grain growth during superplastic deformation. Pseudo single phase alloys can be further divided into: (a) statically recrystallized and, (b) dynamic or continuous recrystallizing alloys. Statically recrystallized alloys are recrystallized to a fine grain size prior to superplastic forming. During the forming process strain induced grain growth occurs resulting in a hardening effect (due to reduced grain boundary sliding and increased dislocation slip). In contrast, dynamically recrystallizing alloys are processed by warm working but left unrecrystallized prior to superplastic deformation. The material dynamically or continuously recrystallizes during superplastic forming. According to Ghosh,[2] no nucleation of new grains is implied but rather a gradual conversion of sub-grain or subcell boundaries into high angle grain boundaries aided by superplastic deformation may occur. Maybe dynamic recovery is a better term for this phenomenon.

This paper reports the results of an investigation of the superplastic flow characteristics of two pseudo-single phase, fine grain, continuously recrystallizing aluminum-lithium alloys: (a) Alcoa 2090-OE16 (2.7% Cu, 2.2% Li, 0.12% Zr), and (b) Alcan 8090-SPF (2.5% Li, 1.2% Cu, 0.7% Mg, 0.12% Zr). Flow stress vs true strain rate curves were obtained experimentally by change strain rate tests at temperatures ranging from 703 K to 783 K (0.75-0.84 T_M) and under a hydrostatic pressure of 400 psi (2.76 MPa). The experimental flow stress versus true strain rate curves are compared to those predicted by superplastic theories and models proposed by Ashby-Verrall,[3] Ball-Hutchinson,[4] and Mukherjee.[5]

PROPOSED MODELS FOR SUPERPLASTIC FLOW

It is generally agreed that superplastic deformation in micrograin superplasticity occurs primarily by grain boundary sliding (GBS) accommodated by either bulk (volume) or grain boundary diffusion or some type of dislocation activity. In fact the proposed accommodation mechanisms for GBS can be divided into two categories: (a) diffusional and, (b) dislocation motion.[6,7] Although there are many proposed models for superplastic deformation,[6,7] only the Ashby-Verrall, Ball-Hutchinson, and Mukherjee models will be investigated in this paper.

Diffusional Accommodation

Ashby and Verrall Model[3]. This theory models superplasticity as a transition region between diffusion accommodated flow at low strain rates and diffusion controlled dislocation climb at high strain rates. It describes a grain switching and rearrangement mechanism involving non-uniform diffusional flow that explains equiaxed grain after superplastic deformation. At low strain rates diffusion accommodated flow accounts for more than 99% of the total strain rate and due to the transient increase in grain boundary area during grain rearrangement a threshold stress is required at very low strain rates. At high strain rates dislocation creep accounts for more than 99% of the total strain rate, and at intermediate strain rates the two mechanisms superimpose on each other. The overall strain rate, $\dot{\epsilon}_{tot}$, will be the sum of the strain rates contributed by each process, i.e.,

$$\dot{\epsilon}_{tot} = \dot{\epsilon}_{diff.acc} + \dot{\epsilon}_{disloc.creep} \tag{1}$$

where

$$\dot{\epsilon}_{diff.acc} = B \frac{\Omega}{kTd^2} \left[\sigma - \frac{0.72\,\Gamma}{d} \right] D_v \left[1 + \frac{3.3\,\delta}{d} \frac{D_B}{D_v} \right] \tag{2}$$

and

$$\dot{\epsilon}_{disloc.creep} = A \frac{D_v Gb}{kT} \left[\frac{\sigma}{G} \right]^n \tag{3}$$

where: $\dot{\epsilon}$ = true tensile strain rate;
      B = 100
      Ω = atomic volume;
      k = Boltzmann's Constant;
      T = absolute temperature;
      σ = flow stress;
      Γ = grain boundary free energy;
      d = grain size;
      δ = thickness of the boundary as a high diffusivity path;
      D_v = bulk diffusion coefficient;
      D_B = boundary diffusion coefficient;
      A,n = empirical constants;
      G = shear modulus.

Dislocation Motion Accommodation

Ball and Hutchinson[4] and Mukherjee Models[5]. Ball and Hutchinson proposed that groups of grains slide as a unit until obstructed by unfavorably oriented grains. The resultant concentration of stress is relieved by dislocation motion in the blocking grains. The dislocations pile up against the opposite grain boundary, and the leading dislocation can then climb into and along the grain boundaries to annihilation sites. The model leads to the following rate equation:

$$\frac{\dot{\epsilon} kT}{D_b Gb} \approx 200 \left(\frac{b}{d} \right)^2 \left(\frac{\sigma}{G} \right)^2 \tag{4}$$

Mukherjee proposed a modification of this model in which grains slide individually rather than in groups and dislocations are produced by ledges and protrusions in the grain boundaries. The rate is then controlled by the climb rate of the lead dislocation into annihilation sites located at grain boundaries and leads to the following rate equation:

$$\frac{\dot{\epsilon}kT}{D_b Gb} \approx 2\left(\frac{b}{d}\right)^2 \left(\frac{\sigma}{G}\right)^2 \qquad [5]$$

The value 200 in Eq. [4] was determined experimentally whereas the value 2 in Eq. [5] was calculated theoretically.

EXPERIMENTAL PROCEDURES

Equipment

The experimental setup is shown in Figure 1. It consists of an INSTRON 3117 Universal Testing Machine controlled by a computer. It is described in a paper by Goforth and Srinivasan.[8] The machine has been modified extensively to make it capable of testing superplastic alloys under constant temperature, strain rate, and hydrostatic pressure. The specimen shown in Figure 2 is heated to the required temperature using a three-zone electrical resistance type (INSTRON 3117) wrap-around furnace with an eleven (11) inch heating zone. The temperature inside the stainless steel retort is controlled by an INSTRON 3120-001 self-adaptive temperature controller. Constant hydrostatic pressure is maintained inside the retort with bottled argon. The specimen is loaded inside the retort in such a way that it is held between the bottom and top pull rods. The load cell is located at the end of the bottom pull rod. The load cell has a capacity of 0 to 300 pounds and is temperature and pressure compensated. The machine is interfaced with personal 286 computer which has the necessary software and hardware to generate signals for testing under conditions of: (a) constant strain rate, (b) change strain rate and, (c) dual strain rate. It also acquires and records the test data.

Test Procedures

The uniaxial tensile testing of superplastic materials involves the control of three parameters: strain rate, temperature, and in the case of aluminum-lithium and some other materials, hydrostatic pressure to prevent cavitation. This paper reports the results of change strain rate tests conducted on the two aluminum-lithium alloys. Detailed results of constant and dual strain rate tests are presented in other papers[8,9,10].

The specimen is loaded inside the retort and then the system is pressurized with argon to a hydrostatic pressure of 400 psi (2.76 MPa). The furnace is preheated and the specimen is heated up to the set temperature by clamping the furnace around the retort. A time period of approximately two and a half hours is usually needed to stabilize the temperature conditions inside the retort. The temperature inside the retort, particularly in the region of the specimen (about four inches), can be monitored and controlled to within 3°C by means of three thermocouples, which are connected to the controller. The temperature data generated during the initial heat up and during the test is stored in the computer in order to generate temperature versus time curves and thus document the temperatures experienced by the specimen during the complete test cycle.

Change Strain Rate Tests. The step change strain rate tests basically followed the procedures described by Hedworth and Stowell[11]. In these tests, the strain rate is increased incrementally

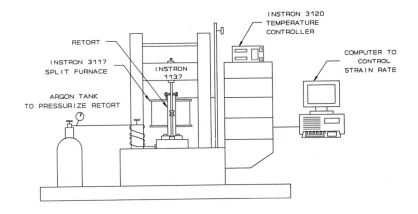

Figure 1 - Experimental Setup

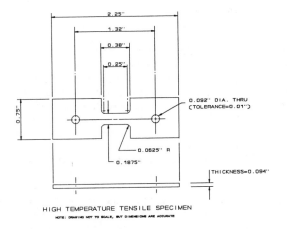

HIGH TEMPERATURE TENSILE SPECIMEN
NOTE: DRAWING NOT TO SCALE, BUT DIMENSIONS ARE ACCURATE

Figure 2 - Specimen Dimensions

from 2×10^{-4} to 1×10^{-2} sec^{-1} allowing the load to stabilize at each strain rate before changing to the next. This sequence is repeated for 3 or 4 cycles until the specimen fails. The load versus displacement curves obtained from each of these tests are then used to develop stress versus strain rate curves for each cycle. The results in this paper utilizes data from the second cycle, which occurs generally in a strain range from 1 to 1.5 and represents that portion of the stress strain curve in which the stress level is increasing or nearly constant. Most practical superplastic forming processes will fall within this strain range, i.e., between approximately 200 and 300% elongation.

RESULTS

The experimental data was fit to a "modified" Ashby-Verrall model by varying B and n. The strain rate associated with diffusional creep is sensitive to the value of B whereas the strain rate associated with dislocation creep (diffusion-controlled dislocation climb) is highly sensitive to the n value. These curves are shown in Figures 3 through 11. Prediction of total strain rate based upon the Ball and Hutchinson model is much slower than that determined experimentally as shown in Figure 12. Mukherjee's model predicts even slower strain rates. In many cases it is impossible to test theories adequately because so much fundamental information, such as grain boundary or interphase diffusivity, is not available. The material property data used to evaluate the models were obtained from a paper by Hamilton[12] and is reproduced in Table I. Linear interpolation was used to obtain values for temperatures not given in Hamilton's paper. The data was originally obtained from Fradin and Rowland[13] and Murr[14]. The grain size used in the analysis was 12.5 μm, based upon measurements of the average grain size using the linear intercept technique.

ANALYSIS OF RESULTS

As can be seen from Figures 3 through 11, the strain rate predicted by the Ashby-Verrall model (B=100) would be several times higher than the experimental values. By reducing B to between 10 and 32, depending upon the temperature, much better agreement between theory and experiment is attained. In general, B increases as temperature increases. It is to be noted that in the lower temperature regimes, B, is fairly close to that predicted by the classical creep model proposed by Nabarro, Herring (B = 14). The value of n varies between 4.8 and 5.2, which is to be expected[15]. The Ashby-Verrall model does not adequately describe superplastic deformation as pointed out by Nix[16]. He states that the sequence of grain shapes proposed by Ashby and Verrall violates the principle of symmetry (two adjacent grains must deform in the same way). Spingarn and Nix[15] have analyzed the suggested diffusional paths in detail and have shown that the Ashby-Verrall process requires grain boundary transport to occur in different directions on opposite sides of the same boundary. They further point out that if grain boundary diffusion is driven by gradients in normal stresses, as usually assumed, then transport along a boundary cannot be different on either side. Nix also points out the necessity for both slip and diffusion to take place in order to satisfy observed grain rotation and texture changes. He argues that these processes can only occur if dislocation motion or slip takes place in some of the grains. This, of course, is different from the diffusion controlled dislocation climb as proposed by Ashby and Verrall. It is thus reasonable to assume that the strain rates under these conditions would be somewhat lower than those predicted by Ashby-Verrall. In fact, they are fairly close to those predicted by the classical creep model (Nabarro, Herring, Coble) where B = 14.

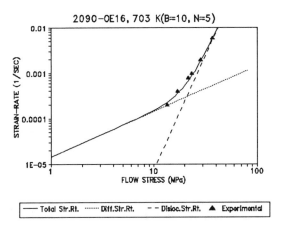

Figure 3 - Comparison of "Modified" Ashby-Verrall Model Calculations With Experimental Data for 2090-OE6 Al at 703°K

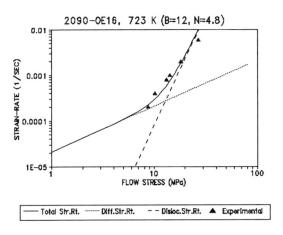

Figure 4 - Comparison of "Modified" Ashby-Verrall Model Calculations With Experimental Data for 2090-OE6 Al at 723°K

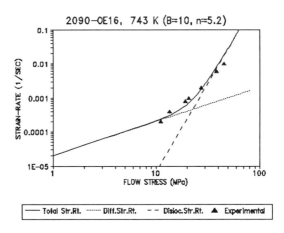

Figure 5 - Comparison of "Modified" Ashby-Verrall Model Calculations With Experimental Data for 2090-OE6 Al at 743°K

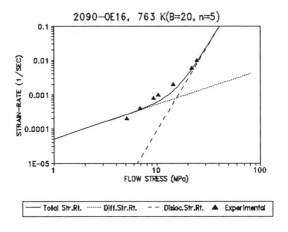

Figure 6 - Comparison of "Modified" Ashby-Verrall Model Calculations With Experimental Data for 2090-OE6 Al at 763°K

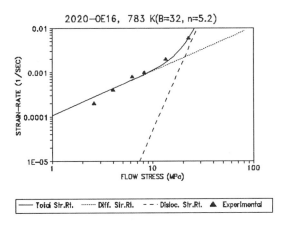

Figure 7 - Comparison of "Modified" Ashby-Verrall Model Calculations With Experimental Data for 2090-OE6 Al at 783°K

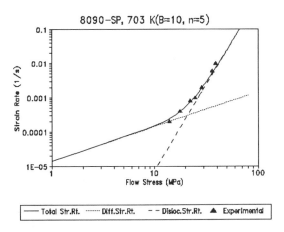

Figure 8 - Comparison of "Modified" Ashby-Verrall Model Calculations With Experimental Data for 8090-SP Al at 703°K

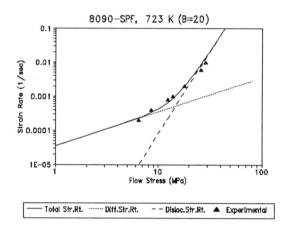

Figure 9 - Comparison of "Modified" Ashby-Verrall Model Calculations With Experimental Data for 8090-SP Al at 723°K

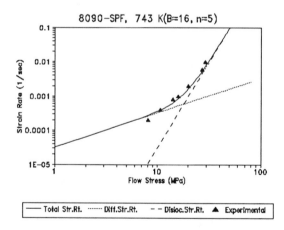

Figure 10 - Comparison of "Modified" Ashby-Verrall Model Calculations With Experimental Data for 8090-SP Al at 743°K

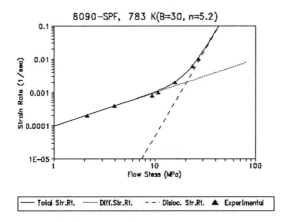

Figure 11 - Comparison of "Modified" Ashby-Verrall Model Calculations With Experimental Data for 8090-SP Al at 783°K

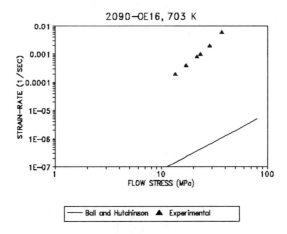

Figure 12 - Comparison of Ball and Hutchinson Model With Experimental Data For 2090-OE16 Al at 703°K.

TABLE I - Data Used in Model Calculations

Temperature (°K)	790	783	763	755	743	723	703
Bulk Diffusivity (cm^2/sec)	8×10^{-10}	6.7×10^{-10}	2.986×10^{-10}	1.5×10^{-10}	1.282×10^{-10}	9.182×10^{-11}	5.545×10^{-11}
Grain Boundary Diffusivity (cm^2/sec)	3×10^{-5}	2.8×10^{-5}	2.23×10^{-5}	2×10^{-5}	1.825×10^{-5}	1.54×10^{-5}	1.24×10^{-5}
Pre-Exponential Constant, A	3.5×10^5	3.3×10^5	2.73×10^5	2.5×10^5	2.19×10^5	1.69×10^5	1.176×10^5
Power-Law Creep Exponent, n	4.8	4.8	4.8	4.8	4.8	4.8	4.8

Atomic Volume, Ω_s: 1.23×10^{-23} (cm^3); Interface Energy, Γ_s: 3.5×10^{-3} (N/cm); Modulus, G_s: 22.5×10^5 (N/cm^2); Burger's Vector, b$_s$: 2.86×10^{-8} (cm); Grain Boundary Width, δ_s: 5×10^{-8} (cm)

The aluminum-lithium alloys studied on this program had similar initial microstructures: a grain size of approximately 3 microns, an aspect ratio of 2:1 (indicating heavy warm working), and a dislocation network of subcells (submicron size) within the grains as revealed by TEM. During the heat-up time (approximately 2 hours to reach a stable, uniform temperature along the sample) and the initial portion of the superplastic deformation cycle, recrystallization occurs and new equiaxed grains grow to 10-15 microns and then seem to stabilize at this grain size for the remainder of the deformation cycle. Dislocation subcell structure within the grains, however, changes dynamically. Some cells seem to grow, others diminish and/or dissipate (possibly into the grain boundaries) during the deformation process. Details of the microstructural evolution is reported in another paper Srinivasan and Goforth[17]. The point is, that there is substantial dislocation activity during the superplastic deformation of these dynamically recrystallizing alloys and this is reflected by the "modified" Ashby-Verrall equation as well as the TEM microstructural evidence.

CONCLUSIONS

1. The Ashby-Verrall model does not adequately describe the superplastic deformation that occurs in the aluminum-lithium alloys studied in this program. The strain rates are much lower than predicted which is reflected by the fact that the coefficient (B) in the diffusional creep portion of the equation is much lower than the 100 predicted by Ashby-Verrall. In fact, it is close to the classical Nabarro, Herring, Coble Creep equation where $B = 14$.

2. The experimental results seem to indicate that grain boundary sliding is accommodated by both diffusion and dislocation slip as proposed by Nix.

3. Results seem to indicate that superplasticity is a transition region between diffusional accommodation and dislocation slip.

REFERENCES

1. John Pilling and Norman Ridley, Superplasticity in Crystalline Solids, (The Institute of Metals, 1989), 9-12.
2. A.K. Ghosh and C. Gandi, Aluminum-Lithium: Development, Applications and Superplasticity, (ASM, 1986), 458.
3. M.F. Ashby and R.A. Verrall, "Diffusion-Accommodated Flow and Superplasticity," Acta Metallurgica, 21 (1973), 149-163.
4. A. Ball and M.M. Hutchinson, Met. Sci. J., 3 (1969) ♭.
5. A.K. Mukherjee, Mater. Sci. Eng., 8 (1971) 83.
6. A. Arieli and A.K. Mukherjee, "Rate-Controlling Deformation Mechanisms in Superplasticity-A Critical Assessment," Metallurgical Transactions A, 13A (1982) 717-732.
7. Jeff W. Edington, "Microstructural Aspects of Superplasticity," Metallurgical Transactions A, 13A (1982) 703715.
8. Ramon E. Goforth and Malur N. Srinivasan, "Testing the Superplastic Flow Characteristics of Advanced Aluminum Alloys," Journal of Testing and Evaluation, Vol. 21, No. 1, (1993) 36-43.
9. R.E. Goforth, et al., "Superplastic Flow Characteristics and Microstructural Analysis of Aluminum-Lithium Alloy 2090-OE16," Superplasticity In Aerospace II, eds.,Terry McNelley and Charles Heikkenen, TMS (1990) 285-302.

10. R.E. Goforth, M. Srinivasan, and N. Chandra, "A Comparative Study of the Superplastic Behavior of Two Dynamically Recrystallizing Aluminum-Lithium Alloys," Superplasticity in Advanced Materials, eds., S. Hori, M. Tokizane and N. Furushiro, JSRS (1991) 145-150.

11. Hedworth, J., Stowell, M.J., "The Measurement of Strain-Rate Sensitivity in Superplastic Alloys," J. Mater. Sci., 6 (1971), 1061-1069.

12. N.E. Paton and C.H. Hamilton, eds., Superplastic Forming of Structural Alloys, (AIME/TMS, 1985), 187.

13. F.Y. Fradin and T.J. Rowland, Applied Physics Letters, 11, (1967) 207.

14. L.E. Murr, Interfacial Phenomenon in Metals and Alloys, Addison Wesley, 1975.

15. J.R. Spingarn and W.D. Nix, "A Model for Creep Based on The Climb of Dislocations at Grain Boundaries," Acta Metallurgica, 27 (1979), 171-192.

16. W.D. Nix, "On Some Fundamental Aspects of Superplastic Flow," WESTEC 1984, Superplastic Forming Symposium, ASM (1984), 3-12.

17. M. Srinivasan and R.E. Goforth, "Microstructural Evaluation of a Dynamically Recrystallizing Superplastic Aluminum-Lithium Alloy," Materials Characterization, Vol. 27, No. 3 (1992).

Constitutive Equations for the Behavior of Superplastic Al-Mg Alloys

T. R. McNelley*, A. A. Salama** and P. N. Kalu*

*Materials Section, Department of Mechanical Engineering
Naval Postgraduate School, Monterey, CA 93943; **Currently with
the Egyptian Military, Cairo, Egypt

Abstract

Constitutive equations for superplastic deformation by slip-accommodated grain boundary sliding and dislocation glide-controlled creep have been combined according to the mantle and core model for independent contributions to the total strain rate. This approach is able to describe the behavior of a fine-grained, processed condition as well as the annealed, coarse-grained condition of an Al-10Mg-0.1Zr (wt. pct.) alloy. Predicted transitions between mechanisms are seen in the experimental data for the processed, fine-grained material. The temperature and strain-rate dependence of the strain-rate sensitivity coefficient and the superplastic ductility are also predicted when grain growth during heating and straining are taken into account.

Advances in Superplasticity and Superplastic Forming
Edited by N. Chandra, H. Garmestani, R.E. Goforth
The Minerals, Metals & Materials Society, 1993

Introduction

The behavior of fine-grained superplastic materials has been described in terms of the additive contributions of independent superplastic and dislocation creep mechanisms [1,2]. Thus,

$$\dot{\epsilon}_t = \dot{\epsilon}_{sp} + \dot{\epsilon}_\perp \tag{1}$$

where $\dot{\epsilon}_t$ is the observed total strain rate, and $\dot{\epsilon}_{sp}$ and $\dot{\epsilon}_\perp$ are the contributions of the superplastic and the dislocation creep mechanisms to the total strain rate, respectively. Superplastic flow is usually regarded to occur by grain boundary sliding with accommodation either by slip or by diffusional flow [1-5]. These processes often are envisioned to occur within the grain boundaries and adjacent, mantle-like regions of the grains [1,2]. The dislocation creep process is then presumed to occur independently in the remaining core regions of the grains and to be unaffected by changes in the grain size. This mantle and core theory predicts that the contribution of the superplastic mechanism will increase as the grain size is reduced and that transitions in deformation mechanism will occur with changes in strain rate, temperature and grain size. However, these two contributions have never been separately evaluated in a single superplastic material thereby providing a direct test of this theory.

Studies of the effects of strain rate, temperature and microstructure on superplastic flow have resulted in several theories of fine-grained superplasticity [3-5]. Phenomenological constitutive equations for the superplastic mechanism, $\dot{\epsilon}_{sp}$, based on these theories have been shown to provide an accurate description of observed behavior for a wide range of fine-grained superplastic materials [2]. The required fine grain size for superplastic response dictates the use of second phases to retard grain growth. Thus, superplastic alloys are often based on eutectic or eutectoid compositions. Alternatively, fine dispersions of second-phase particles have been used to retard grain growth during recrystallization following severe working of some Aluminum alloys [6-11].

Similarly, theories of dislocation creep have resulted in phenomenological equations applicable to pure metals and solid-solution alloys [e.g. 12]. Extension of dislocation creep theories to consider two-phase alloys is difficult unless the dispersion of the second phase is stable. Second-phase dispersions in superplastic materials usually are not stable at superplastic deformation temperatures and so corresponding constitutive relationships for the dislocation creep rate, $\dot{\epsilon}_\perp$, have not been available for the superplastic material. Instead, $\dot{\epsilon}_\perp$ is often assumed to be given by phenomenological equations for single-phase metals and the coefficients are treated as adjustable parameters.

The separate measurement of both $\dot{\epsilon}_{sp}$ and $\dot{\epsilon}_\perp$ for a material would allow the applicability of the assumptions for equation 1 to be assessed. This requires a material capable of being processed to either a fine-grained, superplastic condition or a coarse grained, non-superplastic state. Previous research has shown that superplastic response may be achieved at 300°C in an Al-10Mg-0.1Zr (wt. pct.) alloy. Rolling with controlled interpass anneal (IPA) times at 300°C, a temperature below the β-phase solvus temperature (≈ 365°C), allowed for interaction between evolving deformation structures and β-phase precipitate particles [13-19]. The β-phase particles initiated particle-stimulated nucleation (PSN) during the final stages of the thermomechanical process (TMP) [20,21]. The resultant final grain size of 5μm or less depended on the details of the TMP. The effect of subsequent grain growth on the 300°C superplastic behavior was described in terms of a phenomenological equation for $\dot{\epsilon}_{sp}$ [13].

In a separate study, this same material was processed but then annealed at a temperature of 450°C, resulting in re-solution of the β phase. This was accompanied by extensive grain growth in the absence of the stabilizing effect of the β phase and a resultant grain size of $\approx 50\mu$m. Mechanical test data obtained over a range of temperatures on this coarse grained material was described in terms of a model for glide-control of creep by solute-saturated dislocations [22]. Here, mechanical test data for this same material processed to the superplastic condition were obtained over a similar range of temperature and strain rate. These data are interpreted in terms of equation 1 using a phenomenological equation for superplasticity coupled with the relationship obtained for dislocation creep in this same material.

Experimental Procedures

Material and TMP

The Al-10Mg-0.1Zr (wt. pct.) alloy, in the form of a direct-chill cast ingot, was provided by ALCOA Technical Center, Alcoa Center, Pa. Further details of the composition of this alloy have been given previously [13]. Data for two conditions of the material, as-processed and annealed, were utilized in this study. The essential features of the TMP were: (a), solution heat treatment for 24 hrs at 440°C; (b), upset forging at 440°C to facilitate homogenization, followed by quenching; and (c), heating to 300°C, and rolling, with a controlled IPA time, to a final reduction of $\approx 92\%$ ($\epsilon_{roll} = 2.5$). The second, annealed condition was obtained by heating some of the processed material at 450°C, a temperature above the Mg solvus, to provide a coarse grain size of about 50μm.

Mechanical Testing

Tensile test coupons of gage dimensions 12.7 mm in length, 5.1 mm in width and 2.0 mm in thickness were machined from the as-rolled material with the tensile axes parallel to the rolling direction. The test samples were held in preheated grips and the assembly placed in a furnace attached to an electromechanical machine. Testing was performed at temperatures between 150 and 400°C using constant crosshead speeds providing nominal strain rates between 6.67×10^{-2} sec^{-1} and 6.67×10^{-5} sec^{-1}. All tests were conducted to failure. The same procedures had been employed to study the behavior of the annealed material [22].

Results

The evolution of a superplastic microstructure during the TMP of this alloy has been explained by a model based on the interaction of dislocations with evolving subgrain boundaries and the β-phase precipitate particles [17-19]. With a sufficient IPA time, the equilibrium volume fraction of the β-phase precipitate is attained and a uniform dispersion of spheroidal β particles is achieved [21]. The precipitate particles appear to stabilize subgrain boundary structures resulting in a continuous spread of boundary misorientations up to 15° [17]. The growing β-phase particles may also become sites for PSN resulting in a mixture of recovered structures and recrystallized grains with random, high-angle boundaries [20]. Mean linear intercept measurements ($\approx 1.8\mu$m for the processed material [17-19]) certainly underestimate the true grain size (i.e., the size of regions surrounded by boundaries capable of supporting superplastic deformation mechanisms) for such a microstructure.

The mechanical test data obtained for the as-processed material is summarized in Figure 1(a) as the flow stress at a true strain of 0.1, $\sigma_{0.1}$, versus test temperature. Each curve represents a separate strain rate. The data indicate a normal softening with increasing temperature up to approximately 325°C. Between 325 and 350°C there is an anomalous increase in strength for all strain rates employed. Above this region, softening is again seen up to the highest test temperatures considered. Corresponding activation energy data are shown in Figure 1(b). The maximum superplastic ductility for the processed (as-rolled) material was obtained at ≈ 300°C (Figure 2). Data for the annealed material are included in Figure 2 and indicate a much lower but gradual increasing ductility over this same test temperature range.

The activation energy results corresponding to the data of Figure 1(a) were obtained by plotting $\log \dot{\epsilon}$ versus $1/T$ from isostress lines on the $\sigma_{0.1}$ versus T curves. Values were calculated by the relation

$$Q_{obs,\sigma} = -R \frac{\partial \ln \dot{\epsilon}}{\partial (\frac{1}{T})} \qquad (2)$$

where $Q_{obs,\sigma}$ is the activation energy, R is the gas constant and T is the absolute temperature. The results of these calculations are shown in Figure 1(b). These data reveal two regimes of essentially the same activation energy separated by a transition region. For temperatures below 300°C and stresses from 20 -

47

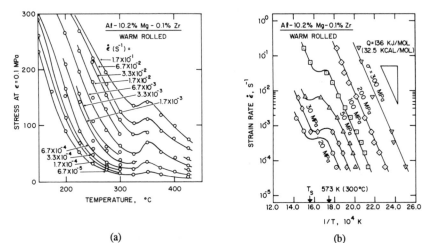

(a)   (b)

Figure 1.   *Flow stress versus temperature (a) and activation energy data (b) for the processed Al-10Mg-0.1Zr alloy.*

300 MPA, $Q_{obs,\sigma}$ = 136 kJ/mol. Above the solvus temperature, 365°C, the same value $Q_{obs,\sigma}$ = 136 kJ/mol was obtained. The transition region from 325 - 350°C, where diminished or even negative activation energy values are apparent, corresponds to the temperature regime of anomalous increase in the flow stress in the data of Figure 1(a). The TMP was conducted at 300°C; microstructural instability resulting from heating to temperatures above this value likely is the origin of these effects. The stabilizing effect of the β particles on the structure would be lost at the Mg returns to solution upon heating in this temperature regime.

In order to assess this correlation, the flow-stress versus temperature and activation energy data for the processed condition were compared to corresponding data for the annealed material. Data for three common strain rates indicate that these two conditions behave identically above 350°C (Figure 3(a); this is also apparent in the ductility data in Figure 2).

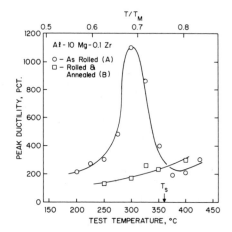

Figure 2.   *Peak ductility versus test temperature for the processed (as-rolled) condition as well as the annealed condition for the Al-10Mg-0.1Zr alloy.*

The fine-grained processed condition is weaker but more ductile than the annealed material below the transition regime and this is also apparent in the comparison of activation energy data shown in Figure 3(b).

48

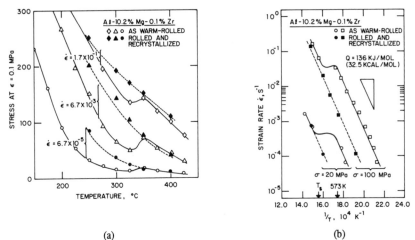

(a)   (b)

Figure 3.   *A comparison of flow stress versus temperature data for the processed and annealed conditions (a), and of activation energy data, (b), for these same conditions of the Al-10Mg-0.1Zr alloy.*

Discussion

The behavior of the processed material can be interpreted in terms of the additive contributions of two independent mechanisms as suggested by equation 1. The activation energy data suggest that the same thermally activated mechanism controls deformation for both the as-processed and annealed conditions at 350°C and above. This mechanism apparently controls the behavior of the annealed material at all temperatures considered. A different mechanism having the same temperature dependence controls the behavior of the processed material below the transition region.

The data for the annealed, coarse-grained material have been interpreted in terms of a dislocation creep mechanism involving glide-controlled motion of solute-saturated dislocations. The value of $Q_{obs,\sigma}$ is consistent with reported values of the activation energy for Mg diffusion in Al (131 kJ/mol) [23,24] and to the reported value of the activation energy for self diffusion in Al (142 kJ/mol) [25]. This suggests that the data for the annealed material can be replotted as diffusion-compensated strain rate versus modulus compensated stress (Figure 4). The solid line, which represents $\dot{\epsilon}_\perp$, is given by

$$\frac{\dot{\epsilon}_\perp}{D} = \frac{K}{\alpha^3} \left(\sinh \alpha\left(\frac{\sigma}{E}\right)\right)^3 \tag{3}$$

where D is the solute (Mg) diffusion coefficient, K is a material constant, α^{-1} is the value of σ/E at the onset of power law breakdown, σ is the stress and E is Young's modulus. Data for the temperature-dependent modulus, E, of pure Al was employed in the absence of data applicable to Al-Mg alloys. For $\sigma/E < \alpha^{-1}$ (below power-law breakdown), equation 3 becomes

$$\frac{\dot{\epsilon}_\perp}{D} \approx K \left(\frac{\sigma}{E}\right)^3 \tag{4}$$

where the strain rate sensitivity coefficient m = 1/n = 0.33. This value has been associated with solute drag on moving dislocations and glide-controlled dislocation creep. The solid line in Figure 4 corresponds to $K = 1.68 \times 10^{21}$ m^{-2} and $\alpha = 900$. A value of $K = 2.65 \times 10^{21}$ m^{-2} was estimated from a model for glide-controlled creep of solid solutions when dislocations become solute saturated.

49

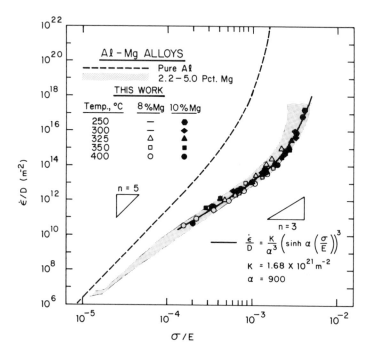

Figure 4.   *Diffusion compensated strain rate $\dot{\varepsilon}/D$ versus modulus compensated stress σ/E for 8 and 10Mg alloys [20].*

The similarity of activation energy values for the processed material (above and below the transition region) to those obtained for the annealed condition suggest the same approach to the data for the processed material. The data for the processed condition (open symbols) are superimposed in Figure 5 upon the data for the annealed condition (filled symbols). For the lowest test temperatures and highest strain rates (large $\dot{\varepsilon}/D$) the behavior of both processing conditions appears to coincide. Also, at test temperatures of 350°C and above the two conditions respond identically. However, for test temperatures ranging from 225 to 325°C the processed material is weaker while m tends toward 0.5. The data for each test temperature in this regime appear to lie on separate curves and comparison of data at constant $\dot{\varepsilon}/D$ for temperatures in this interval reveal the strengthening with increased temperature noted previously. Activation energy values within this temperature interval were lower than those for the data above or below this regime.

The stress and temperature dependence of deformation for many superplastic materials has been shown to follow the phenomenological equation [2]

$$\dot{\varepsilon}_{sp} = A\, D_{eff}^{*} \left(\frac{1}{d}\right)^{2} \left(\frac{\sigma}{E}\right)^{2} \tag{5}$$

where $\dot{\varepsilon}_{sp}$ is the strain rate for superplastic flow, $A \approx 2 \times 10^{9}$, D_{eff}^{*} is an effective diffusion coefficient, d is the grain size, and E is Young's modulus. The effective diffusion coefficient in equation 2 is given by [2]

50

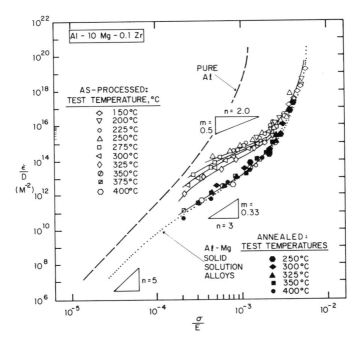

Figure 5. *A comparison between diffusion compensated strain rate $\dot{\epsilon}/D$ versus modulus compensated stress σ/E for both the processed and the annealed condition.*

$$D_{\it eff}^{*} = D_l + (\frac{0.01\pi\delta}{d})\, D_{gb} \tag{6}$$

where D_l is the lattice diffusion coefficient, δ the grain boundary width and D_{gb} is the grain boundary diffusion coefficient.

The behavior of the processed superplastic material may be modeled by combining equations 1, 3 and 5 if grain coarsening upon heating and straining is included:

$$\frac{\dot{\epsilon}_t}{D} = A\, (\frac{1}{d})^2\, (\frac{\sigma}{E})^2 + \frac{K}{\alpha^2}\, (\sinh\, \alpha(\frac{\sigma}{E}))^3 \tag{7}$$

where

$$d = d(\epsilon,\dot{\epsilon},T) \tag{8}$$

and the coefficients have been defined previously. In applying equation 7, it was assumed that $D = D_l$, the lattice diffusion coefficient for Al, because observed activation energies were nearly equal to that for lattice diffusion. The uppermost solid curve in Figure 6 represents the behavior of equation 7 for a grain size $d = 2\mu m$. A predicted transition from dislocation glide-controlled deformation to the superplastic mechanism at $\dot{\epsilon}/D \approx 10^{16}$ m^{-2} is apparent in the experimental data. An additional curve was calculated based on the reported grain growth data, i.e. $d(\epsilon,\dot{\epsilon},T)$, obtained from testing of this

51

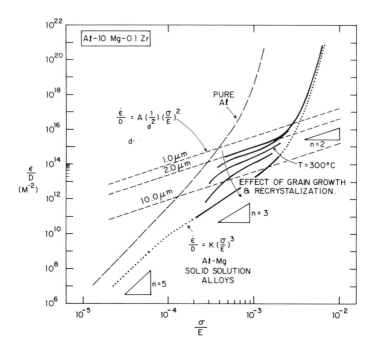

Figure 6. *The predicted behavior of the processed Al-10Mg-0.1Zr alloy when grain growth during heating and straining are taken into account. Compare to Figure 5.*

material at 300°C [13]. In this intermediate temperature regime, the effects of grain growth appear to account for the temperature and strain-rate dependence of the strain rate sensitivity coefficient and thus the ductility of the processed material. The lowest solid curve applies to grain size d $\geq$ 50μm. Grain growth in the processed material upon heating above the solvus temperature results in a second transition in behavior with a return to dislocation-glide controlled deformation.

Conclusions

The behavior of the processed material may be interpreted in terms of independent contributions from "mantle" and "core" mechanisms. This is evident in the close correspondence between the model and the experimental data for this Al-Mg alloy. At large values of $\dot{\epsilon}/D$, the similar strengths of the processed and annealed conditions suggests that the dislocation mechanism is does not depend on grain size.

The onset of grain boundary sliding and superplastic response in the processed material as temperature is increased is predicted by this model. The relatively low temperature range (ca. 300°C) for superplastic response in this material results from the high strength associated with Mg atom - dislocation interaction in the "core". This demonstrates the importance of increasing grain core strength as means of extending the range of superplastic response in Aluminum. Grain growth during straining of the processed material at intermediate temperatures can account for the temperature and strain-rate dependence of the ductility in this regime. Finally, the "mantle" behavior does not appear to be affected by the high Mg content of this alloy.

Acknowledgement

This work was sponsored by the Naval Air Systems Command with Dr. L. E. Sloter as monitor and funding was provided by the Naval Postgraduate School.

References

1. R. C. Gifkins: Metall. Trans., vol. 7A, 1976, p. 1225.
2. O. D. Sherby and J. Wadsworth: in Deformation Processing and Structure (G. Krauss, ed.), ASM, Materials Park, OH, 1982, p. 355.
3. A. Ball and M. M. Hutchinson: Metal Sci. J., vol. 3, 1969, p. 1.
4. T. G. Langdon: Phil. Mag., vol. 22, 1970, p. 689.
5. M. F. Ashby and R. A. Verrall: Acta Metall., vol. 21, 1973, p. 149.
6. B. M. Watts, M. J. Stowell, B. L. Baikie and D. G. E. Owen: Metal Sci. J., 1976, vol.10, pp. 189 and 198.
7. E. Nes: Mater. Sci., 1978, vol.13, pp. 2052.
8. E. Nes: Met. Sci., 1979, vol.13, pp. 211.
9. R. H. Bricknell and J. W. Edington: Metall. Trans. A, vol.10A, 1979, 1257.
10. R. H. Bricknell and J. W. Edington: Acta Metall., 1979, vol. 27, pp. 1303.
11. E. Nes: in Superplasticity (B. Baudelet and M. Suery, eds.), CNRS, Paris, 1985, pp. 7.1.
12. O. D. Sherby and P. M. Burke: Prog. Mater. Sci., vol. 13, 1968, p. 325.
13. E. W. Lee and T. R. McNelley: Mater. Sci. Eng., vol. 93, 1987, p. 45.
14. S. J. Hales and T. R. McNelley: Acta Metall., vol.36, 1988, p. 1229.
15. R. Crooks, S. J. Hales and T. R. McNelley: in Superplasticity and Superplastic Forming (C. H. Hamilton and N. E. Paton, eds.), TMS-AIME, Warrendale, PA, 1988, p. 389.
16. S. J. Hales and T. R. McNelley: in Superplasticity in Aerospace (H.C.Heikkenen and T.R.McNelley, eds.), TMS-AIME, Warrendale, PA, 1988, p. 61.
17. S. J. Hales, T. R. McNelley and H. J. McQueen: Metall. Trans. A, vol. 22A, 1991, P. 1037.
18. S. J. Hales, T. R. McNelley and R. Crooks: in Recrystallization '90 (T. Chandra, ed.), TMS, Warrendale, PA, 1990, p. 231.
19. T. R. McNelley and S. J. Hales: in Superplasticity in Aerospace II (T. R. McNelley and H. C. Heikkenen, eds.), TMS, Warrendale, PA, 1990, p. 207.
20. R. Crooks, P. N. Kalu and T. R. McNelley: Scri. Metall. et Mater., vol. 25, 1991, p. 1321.
21. T. R. McNelley, R. Crooks, P. N. Kalu and S. A. Rogers: to appear in Mater. Sci. Eng.
22. T. R. McNelley, D. J. Michel and A. A. Salama: Scri. Metall. et Mater., vol 23, 1989, p. 1657.
23. G. Moreau, J. A. Cornet and D. Calais: J. Nucl. Mater., vol. 38, p. 197.
24. S. J. Rothman, et al.: Phys. Stat. Sol. (b), vol. 63, 1974, p. K29.
25. T. S. Lundy and J. F. Murdock: J. Appl. Phys., vol. 33, 1962, p. 1671.

MODELING MICROSTRUCTURAL EVOLUTION AND
THE MECHANICAL RESPONSE OF SUPERPLASTIC MATERIALS*

D. R. Lesuer, C. K. Syn, K. L. Cadwell, C. S. Preuss

Engineering Sciences Division
Lawrence Livermore National Laboratory
Livermore, CA 94550

ABSTRACT

A model has been developed that accounts for grain growth during superplastic flow and its subsequent influence on stress-strain-strain rate behavior. These studies are experimentally based and have involved two different types of superplastic materials - a quasi-single phase metal (Coronze 638) and a microduplex metal (ultrahigh-carbon steel - UHCS). In both materials the kinetics of strain-enhanced grain growth have been studied as a function of strain, strain rate and temperature. An equation for the rate of grain growth has been developed that incorporates the influence of temperature. The evolution of the grain size distribution during superplastic deformation has also been investigated. Our model integrates grain growth laws derived from these studies with two mechanism based, rate dependent constitutive laws to predict the stress-strain-strain rate behavior of materials during superplastic deformation. The influence of grain size distribution and its evolution with strain and strain rate on the stress-strain-strain rate behavior has been represented through the use of distributed parameters. The model can capture the stress-strain-strain rate behavior over a wide range of strains and strain rates with a single set of parameters. Many subtle features of the mechanical response of these materials can be adequately predicted.

*Work performed under the auspices of the U. S. Department of Energy by the Lawrence Livermore National Laboratory under contract No. W-7405-ENG-48.

Advances in Superplasticity and Superplastic Forming
Edited by N. Chandra, H. Garmestani, R.E. Goforth
The Minerals, Metals & Materials Society, 1993

INTRODUCTION

It is generally recognized that superplasticity is very sensitive to material microstructure and testing or forming within an appropriate range of strain rate and temperature. The most important microstructural characteristic is grain size and numerous studies have shown the influence of this parameter on flow stress and ductility. However the fine-grained microstructures that are required for superplasticity are highly susceptible to grain growth which can alter the flow stress or the operating deformation mechanism(s). These changes can, in turn, influence the strain rate and temperature for optimum ductility. For these reasons analysis of superplastic forming can benefit from an understanding of microstructural evolution and material models for superplastic flow should account for microstructural changes and their subsequent influence on deformation mechanisms throughout the deformation history.

Numerous microstructural changes can occur in superplastic materials during deformation including static and strain-enhanced grain growth and dynamic recrystallization. In their most general form material models that account for microstructure and its evolution can be quite complex. Fortunately the most common and important microstructural change is grain growth. This paper describes an experimentally-based study into the kinetics of grain growth and the resulting influence of this microstructural change on the stress-strain-strain rate behavior of two superplastic materials. A material model is presented that integrates the kinetics of grain growth with two mechanism-based constitutive equations. The primary interest has been on applied superplastic forming problems and the need for higher forming rates. Thus this study has focussed on grain boundary sliding and slip creep which are the deformation mechanisms typically encountered in problems involving superplastic flow at higher strain rates.

MICROSTRUCTURAL EVOLUTION

Materials

These studies have involved two superplastic materials with significantly different microstructures - ultrahigh-carbon steel (UHCS) which has a microduplex structure and Coronze 638 which is quasi-single phase. In both these materials the dominant microstructural change during superplastic deformation is grain growth and no dynamic recrystallization was expected at the temperatures and strain rates of interest. The UHCS had the composition Fe-1.8C-1.6Al-1.5Cr-.5Mn and had been thermomechanically processed to produce a matrix of ultrafine, equiaxed ferrite grains (mean linear intercept was approximately 0.74 μm) containing spheroidized carbides. The Coronze 638 (Cu-2.8Al-1.8Si-0.4Co) is a commercially available alloy that consists of essentially pure copper containing a sub-micron size dispersion of CoSi and $CoSi_2$ particles. The material was received as a rolled and repeatedly annealed sheet with mean linear intercept grain size equal to 1.9 microns. Further details of the experimental plan have been reported elsewhere[1].

Strain-enhanced grain growth.

Fig. 1 shows the influence of superplastic deformation at 750°C and a true strain rate of .001s^{-1} on the microstructure of UHCS. Fig. 1(a) was taken in the grip section of the sample and was thus exposed to the testing temperature without plastic deformation. Figs. 1(b) and 1(c) were taken in the gage section of samples deformed to true strains of 0.92 and 1.42 respectively. A comparison of the figures shows strain-

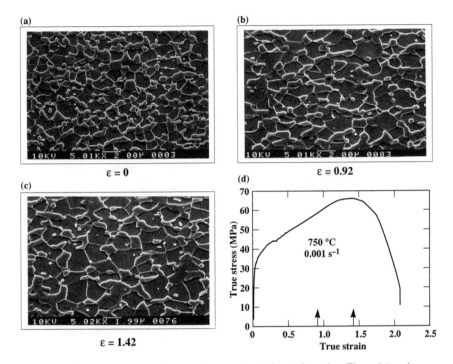

(a)   ε = 0

(b)   ε = 0.92

(c)   ε = 1.42

(d)   750 °C   0.001 s⁻¹

Figure 1. Microstructure of ultra-high carbon steel before deformation (Figure 1a), and after superplastic deformation to a true strain of .92 (Figure 1b) and 1.42 (Figure 1c) respecitvely. Microstructure consists of ferrite grains and iron carbide particles. Superplastic deformation has caused growth in both the ferrite grains and the carbide particles. The true stress-true strain behavior for this material is shown in Figure 1(d). Strains at which the photomicrographs were taken are indicated on the plot.

enhanced grain growth in that the ferrite grains have grown. The carbide particles have also coarsened and the size of the ferrite grains appears to be determined by the intercarbide spacing. This suggests that the kinetics of grain growth are determined by the kinetics of carbide coarsening. The stress-strain curve in Fig. 1(d) shows the importance of this grain growth on the deformation behavior of UHCS - increasing the grain size from its initial size (0.74μm) to the size at a strain of 1.42 (1.48μm) has raised the flow stress from 35 MPa (5 ksi) to over 62 MPa (9 ksi). Thus grain growth has produced significant hardening and the grain size is an important parameter for characterizing the current mechanical state of the material.

In these studies static annealing grain growth (normal grain growth) and strain-enhanced grain growth are assumed to be additive. Thus the kinetics of grain growth can be expressed as

$$\frac{\dot{d}}{d_o} = \frac{\dot{d}_a}{d_o} + \frac{\dot{d}_{se}}{d_o} \tag{1}$$

where $\dot{d}$ is the total rate of grain growth, $\dot{d}_a$ is the grain growth rate due to static annealing, $\dot{d}_{se}$ is the grain growth rate due to strain and d_o is the initial grain size prior to deformation or exposure to elevated temperature.

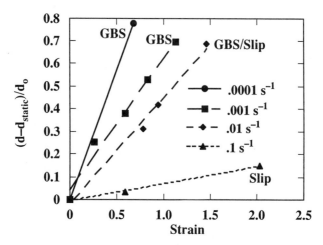

Figure 2. Normalized strain-enhanced grain growth versus strain for Coronze superplastically deformed at four true strain rates. Dominant deformation mechanism is noted for each strain rate.

The grain structure in the gage section of samples is the result of both static and strain-enhanced grain growth. On the other hand the grain structure in the grip is the result of static grain growth only. The strain-enhanced grain growth was taken as the difference in mean linear intercept grain size between measurements taken in the gage and grip sections of the sample. We used this procedure to determine the normalized strain-enhanced grain growth response for Coronze. Wilkinson and Caceres have obtained data for this material.[2] The present studies have obtained data at higher strain rates. The strain-enhanced grain growth for the Coronze is plotted as a function of true strain in Fig. 2 for tests conducted at 550°C and four strain rates. Results have been normalized by the initial grain size. The tests at the three slowest strain rates were in the region in which grain boundary sliding (GBS) is the dominant deformation mechanism. The test at the highest strain rate was in the region where the dominant deformation mechanism was slip creep. For the three slowest strain rates the normalized strain-enhanced grain growth was found to have a linear dependence on strain and a power-law dependence on strain rate. These results are consistent with the observations of Caceres and Wilkinson on the Coronze alloy.[2] For the highest strain rate the grain growth data in Fig. 2 had a much smaller slope. The reason for this will be discussed in the following paragraphs.

The normalized strain-enhanced grain growth rate (with respect to time) can be calculated from the data in Fig. 2 by multiplying the slopes of individual lines by the strain rate for that test. These grain growth rates have been calculated and added to a figure previously reported by Wilkinson and Caceres[2] which shows a log-log plot of normalized grain growth rate versus strain rate. The results are shown in Fig. 3. The plot is quite significant since it shows data for both quasi-single phase and microduplex materials, different homologous temperatures and a range of starting grain sizes. It is clear that a common equation can describe the grain growth behavior of a number of different materials including Coronze. The curve has three distinct regions. In one region the normalized grain growth rate is a power law function of strain rate; at higher or lower strain rates the normalized grain growth rate reaches a limiting value which is independent of strain rate. These regions are shown schematically in the inset for Fig. 3. The region at the lowest strain rate is the result of static grain growth, whereas the regions at the intermediate and high strain rates represent the result of strain-enhanced grain growth. It is reasonable to assume that the highest grain growth rate represents

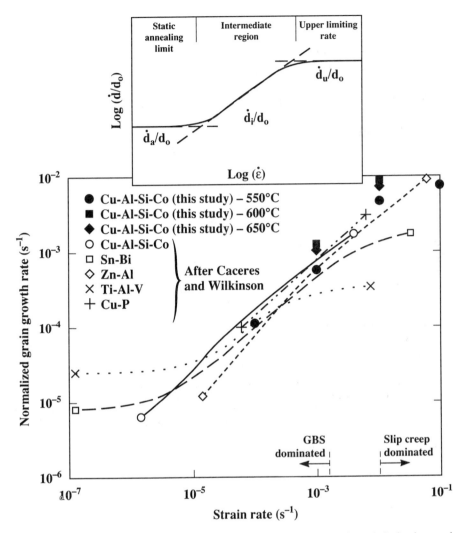

Figure 3. Normalized grain growth rate versus strain rate for a number of quasi-single phase and microduplex superplastic materials. Plot is from the work of Caceres and Wilkinson[2]. Data from this study for Coronze at 550°C, 600°C and 650°C has been added to the plot. The strain rates over which there is a transistion in deformation mechanism from GBS to slip creep are indicated in the figure. The inset shows the three different regions for the curve.

a limiting rate determined by the kinetics of grain boundary migration. The curve shown in Fig. 3 can be described by the following equation

$$\frac{\dot{d}}{d_o} = \frac{\dot{d}_a}{d_o} + \frac{1}{d_o}\left(\frac{\dot{d}_i \, \dot{d}_u}{\dot{d}_i + \dot{d}_u}\right) \tag{2}$$

where $\dot{d}_i$ is the grain growth rate at intermediate strain rates and $\dot{d}_u$ is the upper limiting grain growth rate. The intermediate region has a power law dependence on strain rate which produces the following empirical

59

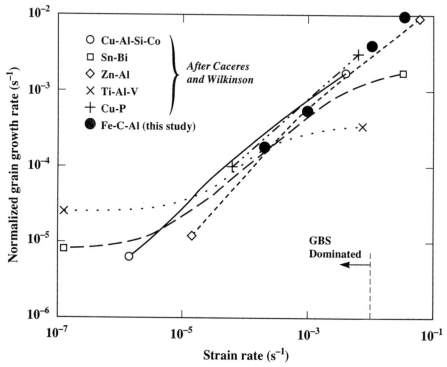

Figure 4. Normalized grain growth rate versus strain rate for a number of quasi-single phase and microduplex superplastic materials. Plot is from the work of Caceres and Wilkinson[3]. Data from this study for the superplastic deformation of UHCS at 750°C has been added to the plot. Strain rate below which GBS is the primary deformation mechanism is indicated on the figure.

expression for the normalized rate of grain growth

$$\frac{\dot{d}}{d_o} = \frac{\dot{d}_a}{d_o} + \frac{1}{d_o}\left(\frac{\lambda_o \dot{\epsilon}^n \dot{d}_u}{\lambda_o \dot{\epsilon}^n + \dot{d}_u}\right) \tag{3}$$

where λ_o and n are constants.

The strain rates at which there is a shift in the operating deformation mechanisms (from GBS to slip creep) are also shown in Fig. 3. It is important to note that the grain growth rates for the three slowest strain rates appear to have a power law dependence on strain rate. The grain growth rate for the highest strain rate, however, shows a substantially smaller increase with increasing strain rate than the grain growth rates at the lower strain rates. This transistion occurs at about the same strain rate as the transistion to slip creep dominated deformation. The obvious implication is that the loss of GBS as a deformation mechanism has reduced the contributions of strain-enhanced grain growth to the total grain growth rate. Several mechanisms have been proposed to explain strain-enhanced grain growth.[3,4,5,6] All of these mechanisms result from grain boundary sliding or grain switching events. It is reasonable to assume that strain-enhanced grain growth will exist only if these mechanisms provide significant contributions to the total strain. Thus contributions from strain-enhanced grain growth can be limited by a loss of superplastic flow or the limiting grain growth rates defined by the rates of grain boundary migration.

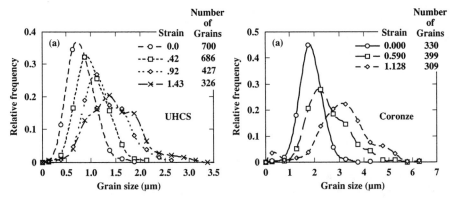

Figure 5. Grain size distributions for superplastically deformed UHCS (5a) and Coronze (5b) for several strain levels. Measurements are based on samples that were deformed to the indicated strains at temperatures of 750°C for UHCS and 550°C for Coronze. The strain rate was .001s⁻¹. The grain size distribution broadens with strain increasing the possiblility of multiple deformation mechanisms operating within the sample.

Identical procedures were used to determine the normalized strain-enhanced grain growth rate for UHCS during superplastic deformation at 750°C. Results are presented in Fig. 4 and fall within the range of grain growth rates for other materials represented on the plot. For UHCS no upper limit is reached on grain growth rate over the strain rates studied.

The distribution of grain sizes in these materials and how they change with strain and strain rate during superplastic deformation has also been examined. Typical results for UHCS and Coronze are shown in Fig. 5. The initial distribution (before deformation) has a log normal appearance. With deformation the mean value of the distribution increases and the distribution broadens. As we will see in the following sections this alteration in the distribution of grain sizes can have a significant impact on the deformation mechanisms operating in the material and the resulting stress-strain-strain rate behavior.

<u>Temperature dependence of strain-enhanced grain growth.</u>

<u>Theory.</u> A general extension of equation (3) which accounts for the temperature dependence of grain growth can be developed assuming different temperature dependencies for the three processes in equation (2). The temperature dependence of normal grain growth kinetics has been studied and equations developed (see for example reference 7). The primary interest in this study is strain-enhanced grain growth and thus the intermediate and upper rate processes represented in equation (2). The temperature dependence of these processes can be represented as follows

$$\dot{d}_i = \lambda_o \dot{\varepsilon}^n \exp\left(\frac{-Q_i}{RT}\right) \tag{4a}$$

$$\dot{d}_u = \left(\dot{d}_u\right)_o \exp\left(\frac{-Q_u}{RT}\right) \tag{4b}$$

where Q_i and Q_u are activation energies for the intermediate and upper regions respectively and λ_o and $(\dot{d}_u)_o$ are constants. Combining these expressions yields a general equation for the temperature dependence of strain-enhanced grain growth.

$$\frac{\dot{d}_{se}}{d_o} = \frac{1}{d_o} \left(\frac{\lambda_o \dot{\varepsilon}^n (\dot{d}_u)_o \exp\left(\frac{-(Q_i + Q_u)}{RT}\right)}{\lambda_o \dot{\varepsilon}^n \exp\left(\frac{-Q_i}{RT}\right) + (\dot{d}_u)_o \exp\left(\frac{-Q_u}{RT}\right)} \right) \tag{5}$$

Experimental Evaluation. The temperature dependence of strain-enhanced grain growth for the Coronze alloy was experimentally evaluated at 600°C and 650°C. The resulting grain growth rates have been added to the plot in Fig. 3 and appear to fall within the range of strain-enhanced grain growth rates for other materials. These results suggest that strain-enhanced grain growth for the Coronze alloy is independent of temperature and Q_i for this material is zero. As mentioned in the previous section the limiting grain growth rate at high strain rate is probably controlled by the rate of grain boundary migration. We therefore assume that a reasonable value for Q_u is the activation energy for grain boundary diffusion. The calculated grain growth rates that are predicted by equation (5) are plotted as a function of strain rate in Fig. 6. Calculations are shown for three temperatures (450°C, 550°C and 650°C) and the parameters used in equation (5) are given in Table I. The calculated grain growth rates show good agreement with rates derived from experimental data. Because the strain-enhanced grain growth is independent of temperature in the intermediate region, at very high temperatures (higher than the temperature studied here) the contribution of static annealing to the total grain growth rate could be significantly higher than the contribution from strain-enhanced gtrain growth. In Table I the parameters for UHCS are also given. Both Coronze and UHCS have very similar strain rate exponents (n) and values for the constant λ.

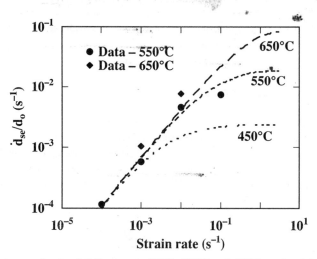

Figure 6. Grain growth rates for Coronze at 450°C, 550°C and 650°C predicted by equation (5). Calculations are based on the parameters given in Table I. Experimental data at 550°C and 650°C are given.

Table I

Parameters Used in Equation (5)

	d_o (microns)	λ_o $((\mu m/s)s^n)$	$(\dot{d}_u)_o$ $(\mu m/s)$	n	Q_i	Q_u (kJ/mole)
Coronze	1.9	.178	7.98×10^4	.806	0	1.04×10^5*
UHCS	.74	.156	1.92×10^7	.799	0	1.70×10^5†

* activation energy for grain boundary diffusion in pure copper - from reference (8)

† activation energy for grain boundary diffusion in pure iron - from reference (9)

The final grain size that would be obtained for Coronze after tensile testing (to a true strain equal to 1) at a constant strain rate are shown in Fig. 7. Calculations, which were done for three temperatures (450°C, 550°C and 650°C), are based on equation (5) using the parameters given in Table I and thus do not include the effects of static annealing. Results in Fig. 7 show a decreasing final grain size with increasing strain rate and very little grain growth above $10^{-1}s^{-1}$ for all testing temperatures. Grain growth decreases with increasing strain rate (Fig. 7) despite the increasing grain growth rate with increasing strain rate (Fig. 6). This occurs because the strain rate exponent (n) in equation (5) is less than one. The total amount of grain growth is very sensitive to the value of n.

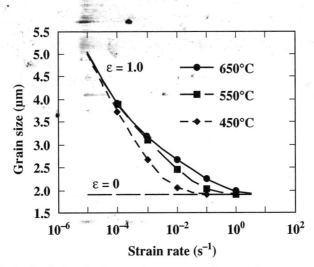

Figure 7.   Calculated grain sizes for Coronze after constant strain rate testing to a true strain of 1 at the indicated strain rates.

MECHANICAL RESPONSE

Modeling with a mean grain size.

Theory. Two rate dependent constitutive equations have been used for GBS and slip creep. These equations can be expressed as (6) and (7) respectively[10]

$$\dot{\varepsilon}_{gbs} = A_{gbs}\left(\frac{b}{d}\right)^3\left(\frac{D_{gb}}{b^2}\right)\left(\frac{\sigma}{E}\right)^2 \tag{6}$$

$$\dot{\varepsilon}_{slip} = A_s\left(\frac{\lambda}{b}\right)^3\left(\frac{D_L}{b^2}\right)\left(\frac{\sigma}{E}\right)^8 \tag{7}$$

where $\dot{\varepsilon}_{gbs}$ and $\dot{\varepsilon}_{slip}$ are the strain rates for grain boundary sliding and slip creep, respectively, σ is the stress, A_{gbs} and A_s are constants, λ is the minimum barrier spacing governing slip creep (typically the interparticle spacing or the grain size), d is the grain size, D_L and D_{gb} are diffusion coefficients, b is the magnitude of the Burger's vector and E is Young's modulus. Since the deformation mechanisms represented by these equations are additive the total strain rate can be represented by

$$\dot{\varepsilon}_{total} = \dot{\varepsilon}_{slip} + \dot{\varepsilon}_{gbs} \tag{8}$$

The mean linear intercept grain size is typically used for the grain size term in equations (6) and (7). For fine grain materials deforming in or near the region of GBS, the minimum barrier spacing is the grain size and thus for these studies we have assumed that λ equals d. For these studies the microstructural evolution has ben characterized in terms of this mean linear intercept grain size using equation (9).

$$\frac{d - d_o}{d_o} = k\, \varepsilon\, \dot{\varepsilon}^p \tag{9}$$

where k and p are constants.

The model combined this grain growth equation with constitutive equations (6) and (7) from which the stress-strain response of a material during superplastic deformation was determined.

Experimental Evaluation. Results of the stress-strain calculations for UHCS at three strain rates are shown in Fig. 8. The dominant deformation mechanism for each strain rate is indicated in the figure. Two tests were conducted at strain rates at which the deformation was predominantly by GBS. One of these tests was conducted at a constant true strain rate while the other was conducted at a constant extension rate. Model calculations are compared with experimental data. In both cases the model predicts the hardening behavior of UHCS very well which dramatically illustrates that, when GBS is the dominant deformation mode, the hardening behavior of UHCS is entirely due to grain growth. The third test was conducted at a strain rate

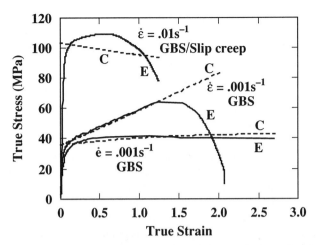

Figure 8. Stress versus strain for the UHCS at 750°C and three strain rates. Dominant deformation mechanism for each strain rate is indicated in the figure. The calculated stress-strain response (indicated with a C) was obtained by using equations (6), (7), (8) and (9). A mean grain size was used in these calculations to characterize the grain size distribution. The experimental data is indicated with an E.

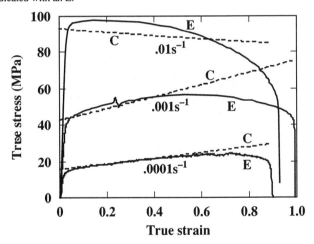

Figure 9. Stress versus strain for the copper alloy at 550°C and three strain rates. The calculated stress-strain response (indicated with a C) was obtained by combining the grain growth equation with the two constitutive laws. A mean grain size was used in these calculations to characterize the grain size distribution (such as shown in Figure 5). The experimental data was taken at a constant true strain rate and is indicated with an E.

at which deformation by GBS and slip creep had significant contributions. Agreement between the calculated results and experimental data is not good. The primary reason for this discrepancy is the strong influence of grain size distribution on stress-strain behavior when two deformation mechanisms have significant contributions. The influence of grain size distribution is discussed in the next section.

Results of the stress-strain calculations for Coronze deformed at three different strain rates are shown in

Fig. 9. General agreement was obtained with experimental data; however two difficiencies are clearly evident. The model does not capture the correct curvature of the stress-strain curves. It also does not predict a maximum in the some stress-strain curves and subsequent softening. The biggest problem with this model is that it characterizes the spectrum of grain sizes actually observed in materials (as illustrated in Fig. 5) with a single average value. We correct for this deficiency in the following section.

Modeling with a grain size distribution.

Theory. For this work the stress-strain-strain rate response was calculated using a distributed parameters approach originally proposed by Ghosh and Raj[11,12,13] in which the stress is calculated from a distribution of grain sizes. In our case this distribution evolves as a function of strain and strain rate. Specifically for our model we assume that during deformation each grain in the material deforms with the same strain rate (which is equal to the strain rate imposed on the material) and thus different size grains will support different stresses. The total stress can be taken as a summation over all grains in the material as

$$\sigma = \Sigma f_i \sigma_i \tag{10}$$

where f_i = volume fraction of grains having a grain size d and σ_i = stress supported by grains with grain size d_i. The values for f_i as a function of grain size were taken from grain size distribution plots. Typical plots are shown in Fig. 5. The stresses for individual grain sizes were calculated for isothermal conditions using equations (11) and (12)

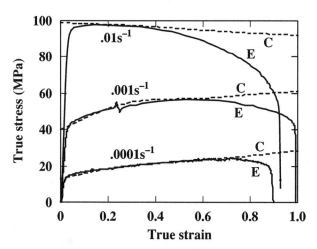

Figure 10. Stress versus strain for the copper alloy at 550°C and three strain rates. The calculated curves (indicated with a C) were obtained from the model using the distributed parameters approach which accounts for grain size distribution, its evolution and multiple deformation mechanisms operating in the material. Experimental data was taken at a constant true strain rate and is indicated with an E. All curves in Figures 10, 11, and 12 were obtained with the one set of parameters shown in Table II.

$$\dot{\varepsilon}_{gbs} = k_{gbs}\sigma_i^{n_{gbs}}d_i^{-q} \tag{11}$$

$$\dot{\varepsilon}_{slip} = k_{slip}\sigma_i^{n_{slip}}d_i^{r} \tag{12}$$

where k_{gbs}, k_{slip}, n_{gbs}, n_{slip}, r and q are constants. An iterative procedure was used to evaluate σ_i and insured that equation (8) was satisfied with the total strain rate equal to the strain rate applied to the material. Using this procedure different grains in the distribution can deform by different deformation mechanisms and the relative contribution from the different mechanisms can change as the material is deformed.

Experimental Evaluation. The stress-strain behavior for coronze at 550°C predicted by this model is shown in Fig. 10. Similarly the stress-strain rate behavior for coronze at 550°C is shown in Fig. 11. The material constants used in equations (11) and (12) for these calculation are given in Table II. *It is important to recognize that a single set of material constants was used for all calculations shown in Figs. 10 and 11.* The agreement

Table II
Parameters used in Equations (11) and (12) for Coronze

k_{gbs} $(s^{-1}\,(MPa)^{-n}\,(\mu m)^q)$	n_{gbs}	q	k_{slip} $(s^{-1}\,(MPa)^{-n}\,(\mu m)^{-r})$	n_{slip}	r
4.98×10^{-6}	2.0	2.6	4.50×10^{-20}	8.0	3.0

σ_i in MPa, d_i in μm

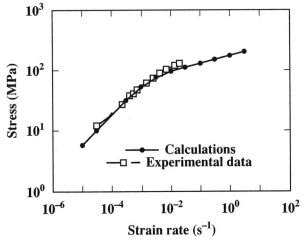

Figure 11. Stress versus strain rate for the copper alloy at 550°C. The calculated curves were obtained from the model using the distributed parameters approach. Experimental data is indicated.

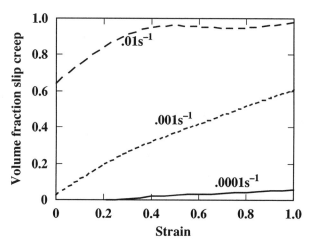

Figure 12. The volume fraction of deformation from slip creep as a function of strain during tensile testing of the copper alloy at 550°C and three different strain rates.

between calculations and experimental data is quite remarkable in that the entire stress-strain curve can be predicted (to the maximum in the stress-strain curve) for three strain rates covering two orders of magnitude in a strain rate range where two deformation mechanisms have significant contributions. The stress-strain rate behavior can be reasonably predicted over the limits of the experimental data (3 orders of magnitude). The stress-strain behavior shown in Fig. 10 is complicated by cavities that nucleate and grow during superplastic deformation of Coronze. Eventually these cavities will interlink to cause fracture. The maximum in the stress-strain curve shown in Fig. 10 and subsequent softening in this curve are probably caused by cavity interlinkage. In other materials that do not form extensive cavities, this maximum could be due to necking (flow localization) of the sample. Fig. 11 shows that the model predicts that the transistion between GBS dominated deformation and slip creep dominated deformation takes place over about one and a half orders of magnitude in strain rate. This observation is consistent with experimental stress-strain rate data in a number of superplastic materials. Fig. 12 shows the volume fraction of slip creep predicted from the model during tensile testing of Coronze at 550°C. Results are shown for the three strain rates reported in Fig. 10. It is clear that substantial changes in the relative contributions to the strain from the deformation mechanisms can occur during testing for strain rates at or near the transistion between regions that are dominated by GBS or slip creep. From Figs. 10 and 12 it is apparent that the curvature shown in the stress-strain curves in Fig. 10 is due to the increasing contribution of slip creep to the deformation as grain growth occurs. If the contribution of grain boundary sliding to the deformation was constant throughout the deformation history then the stress would increase in a linear manner with strain (since from equation (9) grain size is a linear function of strain). Many true stress-true strain curves reported in the literature (e.g. ref. 13) also show, for superplastic deformation under constant strain rate conditions, a linear dependence of stress on strain up to a stress maximum.

CONCLUSIONS

1. The dependence of grain growth rate on strain rate for superplastic UHCS and Coronze falls on a master curve as originally proposed by Wilkinson and Caceres. In the copper alloy the transistion in grain growth rate from a power law dependence on strain rate to an upper limiting rate occurs at the transistion from GBS dominated behavior to slip creep dominated behavior. The transistion to an upper limiting rate ($\dot{d}_u$ in Fig 3) can occur because of a loss of superplastic flow or a limiting grain growth rate defined by grain boundary migration. For UHCS, within the strain rates studied, no upper limit was found to the grain growth rate.

2. An equation describing the temperature dependence of the strain-enhanced grain growth rate has been developed. The equation predicts grain growth rates that agree well with experimental data. For Coronze strain-enhanced grain growth appears to be independent of temperature in the intermediate region. In the high strain rate region the strain-enhanced grain growth rate appears to have an activation energy equal to the activation energy for grain boundary diffusion .

3. A simple material model containing an empirical grain growth law and mechanism-based constitutive equations can provide a reasonable representation of hardening during superplastic deformation. The hardening behavior of UHCS is entirely due to grain growth when GBS is the dominant deformation mechanism.

4. The model can be substantially enhanced by incorporating the influence of grain size distribution (and the evolution of this distribution with strain and strain rate) using constitutive equations containing distributed parameters.

5. The model containing constitutive equations with distributed parameters can capture the stress-strain-strain rate behavior over a wide range of strains and strain rates with a single set of parameters. Many subtle features of the mechanical response of these materials can be adequately predicted.

ACKNOWLEDGEMENTS

We are indebted to Oleg Sherby (Stanford University) and Amiya Mukherjee (University of California at Davis) for helpful discussions on superplasticity. We are also indebted to Jack Crane (Olin Corporation) for providing the Coronze 638 and Oleg Sherby for providing the superplastic ultrahigh-carbon steel. This work was supported by the Laboratory Directed Research and Development Program at Lawrence Livermore National Laboratory.

REFERENCES

1. D. R. Lesuer, C. K. Syn, K. L. Cadwell and S. C. Mance, "Microstructural Change and Its Influence on Stress-Strain Behavior of Superplastic Materials", Superplasticity in Advanced Materials, S. Hori, M. Tokizane, and N. Furushiro, eds, Osaka, 139–144 (1991).

2. D. S. Wilkinson and C. H. Caceres, "An Evaluation of Available Data for Strain-Enhanced grain Growth During Superplastic Flow", J. Mater.Sci. Lett. **3**, 395–399 (1984).

3. M. A. Clark and T. H. Alden, "Deformation Enhanced Grain growth in a Superplastic Sn-1% Bi Alloy", Acta Metall., **21**, 1195–1206 (1973).

4. D. S. Wilkinson and C. H. Caceres, "On the Mechanism of Strain-Enhanced grain Growth During Superplastic Deformation", Acta Metall., **32**, 9, 1335–1345 (1984).

5. K. Holm, J. D. Embury and G. R. Purdy, "The Structure and Properties of Microduplex Zr-Nb Alloys", Acta Metall., **25**, 1191–1200 (1977).

6. D. J. Sherwood and C. H. Hamilton, "A Mechanism for Deformation-Enhanced grain Growth in Single Phase Materials", Scripta Metall. et Mat., **25**, 2873–2878, (1991).

7. P. Cotterill and P. R. Mould, Recrystallization and Grain Growth in Metals, New York, NY: John Wiley & Sons, 279 (1976).

8. M. F. Ashby, "A First Report on Deformation-Mechanism Maps", Acta Metall., **20**, 887–897 (1972).

9. B. Walser and O. D. Sherby, "Mechanical Behavior of superplastic Ultrahigh Carbon Steels at Elevated Temperature", Met. Trans. A, **10A**, 1461–1471 (1979).

10. O. D. Sherby and J Wadsworth, "Superplasticity - Recent Advance and Future Directions", Progress in Materials Science, **39**, 169–221 (1989).

11. A. K. Ghosh and R. Raj, "Grain Size Distribution Effects in Superplasticity", Acta Metall., **29**, 607–616 (1981).

12. R. Raj and A. K. Ghosh, "Micromechanical Modeling of Creep Using Distributed Parameters", Acta Metall., **29**, 283–292 (1981).

13. A. K. Ghosh and R. Raj, "The Evolution of Grain Size Distribution During Superplastic Deformation", Proceedings Int. Conf. on Superplasticity, Grenoble, 11.1–11.19 (1985).

SUPERPLASTICITY IN COMPOSITES
AND   HIGH STRAIN-RATE SUPERPLASTICITY

High Rate Superplasticity in Al_2O_3-ZrO_2-Al_2TiO_5 Ceramics

J.A. Payne and J. Pilling

College of Engineering
Department of Metallurgical and Materials Engineering
Michigan Technological University
Houghton, MI 49931

Abstract

This investigation examines the true stress-true strain behavior of Al_2O_3-ZrO_2-Al_2TiO_5 blends having Al_2TiO_5 contents of 13.9 to 20.1 volume percent which were tested in compression at constant strain rate. Although the mixtures exhibited strain hardening, the flow stress was shown to be low to moderate at intermediate to high strain rates ($\approx$ 10 MPa @ 2×10^{-4} sec^{-1}, 26-38 MPa @ 2×10^{-3} sec^{-1}, and 125+ MPa @ 2×10^{-2} sec^{-1}). Strain rate jump tests indicate that the strain rate sensitivity appears to peak and is quite high (0.7+) at intermediate strain rates (10^{-3} sec^{-1}). Microstructural studies suggest that the ternary blends are resistant to grain growth and appear to be more sensitive to time at temperature than to the imposed strain rate.

Advances in Superplasticity and Superplastic Forming
Edited by N. Chandra, H. Garmestani, R.E. Goforth
The Minerals, Metals & Materials Society, 1993

Introduction

The discovery of superplastic flow in fine grained tetragonal zirconia polycrystals by Wakai [1] in 1986 has lead to considerable research in the generality of this phenomenon in oxide and non-oxide ceramics. Typical materials which have been shown to deform superplastically are: Al_2O_3, ZrO_2, ZrO_2/Al_2O_3, ZrO_2 /mullite, and various silicon nitride blends [2].

These materials exhibit classic superplastic behavior in that they deform according to:

$$\dot{\epsilon} = A \frac{\sigma^n}{d^p} \qquad \text{(EQ 1)}$$

where $\dot{\epsilon}$ is the strain rate, σ is the flow stress, d is the grain size, n is the stress exponent (n=1/m where m is the strain rate sensitivity), p is the grain size exponent, and A is a temperature dependent, diffusion related coefficient.

It is generally stated that to deform superplastically, ceramics should have an initial grain size of less than 10 microns [3]. In a manner exactly analogous to metals, grain growth can lead to the loss of superplastic behavior. It has been reported that an increase in grain size from 0.42 to 3 μm results in at least a five fold increase in flow stress for yittra stabilized zirconia tested in compression at 1450°C [4]. This illustrates the large effect of grain growth on the superplastic properties of ceramics.

Ceramics would see greater commercial utilization if component costs could be reduced. As much as 80% of the final component price can be due to machining [5]. Near net shape forming of ceramic components could be a viable alternative to this dilemma if deformation rates could be increased.

The advent of nanophase powders presents the possibility for "normal" deformation to be conducted at decreased temperatures or, if the structure can be stabilized against grain growth, increased deformation rates at "normal" deformation temperatures. The basic premise of the current research is that a fine, stable mixture of three mutually insoluble phases could inhibit grain growth sufficiently to yield a microstructure with enhanced deformation capabilities, both in terms of deformation rate and extent.

This paper reports the deformation behavior observed for ZrO_2-Al_2O_3-Al_2TiO_5 blends at strain rates as high as 2×10^{-2} sec^{-1}. It is shown that the flow stress for these ternary blends is less than that for binary mixtures at approximately equivalent conditions. Although particle coarsening does occur during deformation, it appears to be minimal.

Experimental Procedure

The experimental activity associated with this study involves constant strain rate compression testing of ternary ceramic mixtures to predetermined strains. Strain rate jump tests were also performed along with a particle coarsening study.

Specimen Preparation

Although the specific details of specimen preparation are currently proprietary; in general, samples were produced as follows. Powder mixtures were prepared by blending Al_2O_3, 3 mol% Y_2O_3 stabilized ZrO_2, and TiO_2. The mixed powders were then uniaxially pressed to shape followed by cold isostatic pressing. All samples were then sintered in air. After sintering the specimens consisted of a ternary mixture of Al_2O_3, ZrO_2, and Al_2TiO_5. Specimen geometry was a circular cylinder typically 10 mm in height and 18 mm in diameter. Since this system produces Al_2TiO_5 in-situ by a reaction between Al_2O_3 and TiO_2, all reference to the alloys used will be in terms of the volume percent of the phases present with particular attention given to the Al_2TiO_5 content.

Testing Procedure

Compression testing was conducted in a Centorr M60 vacuum furnace using molybdenum compression platens lubricated with BN[1]. All deformation studies were performed at 1500°C under constant true strain rate conditions using an Instron Model 4206. To maintain constant strain rate, experiments were monitored/controlled by a PC which subtracted the machine stiffness (determined for the test conditions of interest) from the crosshead displacement. Assuming constant sample volume, this gave an estimation of the instantaneous specimen height, h_i. Once h_i was known the crosshead velocity, V_i, was adjusted to maintain constant strain rate per EQ 2.

$$V_i = \dot{\varepsilon} h_i \qquad \text{(EQ 2)}$$

During strain rate jump testing the specimen was prestrained to a true strain of 0.2 at a strain rate of 4×10^{-4} sec[-1]. The strain rate was then decreased to 5×10^{-5} sec[-1] and the specimen allowed to reach equilibrium, whereupon the strain rate was systematically increased by a factor of two. Regression analysis was used to fit a polynomial to the log(σ)-log($\dot{\varepsilon}$) data. Differentiation of this polynomial allowed the determination of the strain rate sensitivity, m, as a function of the strain rate. Strain rate jump tests were also conducted using a constant strain rate.

Temperature was measured on the molybdenum platens, just under the specimen, using a Model 10 optical pyrometer manufactured by Accufiber[2]. This unit allowed temperature measurements to be corrected for the emissivity of the platens. It was felt that measuring the temperature of the platen under the specimen would provide greater consistency than measuring the specimen temperature directly since there would be no change in emissivity caused by changes in composition. There was also the distinct possibility that at large compressions the pyrometer would inadvertently measure the platen temperature while applying the emissivity correction for the specimen. Comparisons between a thermocouple and the pyrometer indicated that the pyrometer was accurate to within ±3°C over the temperature range of interest.

Sample preparation and Microscopic Examination

Samples were characterized for phase fraction with a Scintag XDS 2000 Diffractometer using Cu Kα radiation and a graphite monochromator. Volume fraction was determined using the direct comparison method using as many non-overlapping peaks as possible for each phase. Experiments at Michigan Technological University have shown that this method typically provides relative accuracies of ±5%.

Specimens were sectioned through the center on a plane parallel to the loading axis. After polishing, the samples were thermally etched for 2 hours at 1400°C. Particle size was determined by quantitative metallography [6] using a circular grid and the following formula:

---

1. ZYP Coatings Inc., Oak Ridge, TN.
2. Luxtron Corp., Beaverton, OR.

$$d_\alpha = \frac{1.558nCV_\alpha}{N_\alpha} \qquad \text{(EQ 3)}$$

where: d_α is the particle size of the α phase, n is the number of grids applied, C is the circumference of the grid at the current magnification, V_α is the α volume fraction, N_α is the total number of grid-α intersections, and the constant is a shape factor with assumes non-textured tetrakaidecahedral grains having a log-normal size distribution [7].

Results and Discussion

Three ceramic blends have been investigated to date. Table I summarizes the physical data for the sintered alloys. A goal of this work is to characterize the grain growth, both statically and dynamacially, as a function of ternary composition. Hence, samples of each alloy were deformed to predetermined strains (0.2, 0.5, 0.7, 0.9, and 1.1) at various strain rates. This, when combined with static grain growth experiments, will allow an unambiguous determination of the dynamic grain growth behavior.

Table I: Physical data for ceramic blends.

X-Ray Volume Percent[†]			X-Ray Density (g/cc)	Measured Density (g/cc)[‡]	Percent of Theoretical
Al_2O_3	ZrO_2	Al_2TiO_5			
23.7	62.4	13.9	5.1	4.98	98%
36.3	45.8	17.9	4.78	4.74	99%
41.6	38.4	20.1	4.64	4.57	98%

†. By the Direct Comparison Method.
‡. Using water displacement.

Figure 1 shows the stress-strain behavior observed for the alloys tested at slow to intermediate strain rates. Although there is some scatter in the data, it is clear that all these alloys show strain hardening. As the volume fraction of Al_2TiO_5 increases from 13.9 to 17.9 v/o the flow stress decreases, regardless of strain rate. Though the amount of data is limited, it would appear that a further increase in Al_2TiO_5 to 20.1 v/o does not cause any significant change in the flow stress. To date, one test has been conducted at a strain rate of 2×10^{-2} sec^{-1}. This data, for the 13.9 v/o Al_2TiO_5 alloy, is shown in Figure 2. It should be noted that the specimen did not fail; the test was aborted because the capacity of the load cell was reached.

For easier comparison Table II summarizes the observed behavior. Table III tabulates flow behavior for various alloys and test methods reported in the literature. It should be noted that most of these tests were not conducted at constant strain rate. Thus the strain rate decreased during tension tests and increased for compression tests. Although it is difficult to compare these data due to differences in deformation temperature, imposed strain rate conditions, alloy content, and frictional constraints; in general the flow stress would appear to be decreased for ternary blends, especially at higher strain rates.

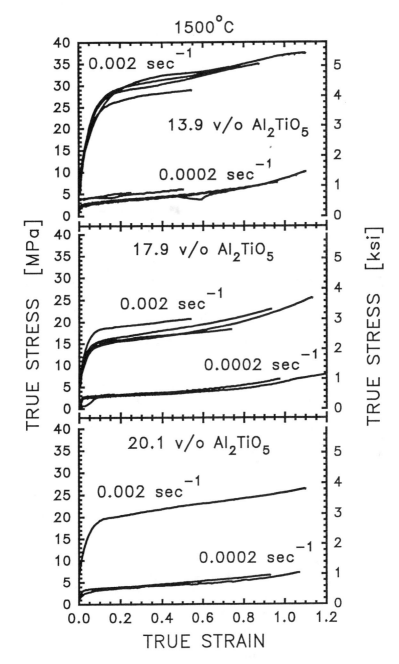

Figure 1. True Stress-True Strain Data for ternary ceramic blends with Al_2TiO_5 contents varying from 13.9 to 20.1 v/o.

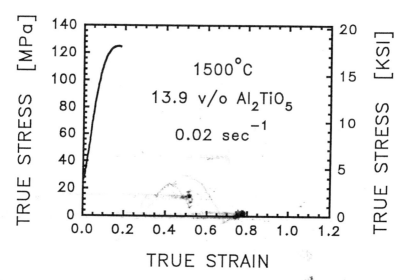

Figure 2. True Stress-True Strain Data for a 13.9 v/o Al_2TiO_5 blend tested at a strain rate of 2×10^{-2} sec^{-1}.

Figure 3 shows the result from strain rate sensitivity determinations. The solid circles represent the flow stress from the jump tests while the open circles are average stress from the constant strain rate tests taken at $\varepsilon = 0.3$. The filled triangles are the slope (m) of the fitted jump test data. At slow strain rates the flow stresses observed for the two methods were similar. At higher strain rates the conventional flow stress was always less than that observed in the jump tests. Although this could be an artifact of the specimen geometry (discussed below) it is more likely caused by microstructural coarsening in the jump test due to the greatly increased time that the specimen is at temperature [8]. Hence, the m values shown are probably skewed toward slightly larger values at higher strain rates.

Figure 4 shows macrophotos of an undeformed and 2 deformed specimens (different strain rates) of the 17.9 v/o blend. These samples are typical of the specimen geometry observed during testing. It should be noted that high strain rate specimens ($\geq 2 \times 10^{-3}$ sec^{-1}) had edge cracks nucleate when the $\varepsilon \approx 0.5$ although the cracks remained small until $\varepsilon \approx 1$. Although Al_2TiO_5 is known to microcrack [12], no strong correlation between edge cracking and the amount of Al_2TiO_5 was noted, perhaps due to the relatively limited range of aluminum titanate content. This cracking may not be detrimental in a closed-die forging since crack healing has been observed during upsetting in these ceramics.

The change in microstructure during testing of a 17.9 v/o Al_2TiO_5 sample is shown in Figure 5. These micrographs are backscattered images showing particle growth observed in samples deformed at 2×10^{-4} sec^{-1} ($\varepsilon = 1.21$) and 2×10^{-3} sec^{-1} ($\varepsilon = 1.13$) versus the as-sintered structure. The white phase is ZrO_2, the gray phase is Al_2TiO_5 and the black phase is Al_2O_3. The phase sizes are shown in Table IV, from which it appears that the ternary blends are effective at minimizing the growth of all phases (the maximum change in particle size observed was approximately 1.4 μm). Since the slower strain rates generally produce larger particle sizes, it would appear that time at temperature is of greater importance than the imposed strain rate during particle growth.

Table II: Tabulated Flow data for the current work at 1500°C

Volume Percent Al_2TiO_5	$\dot{\varepsilon}$ sec^{-1}	Maximum Flow Stress MPa	ε @ Max σ
13.9		10.1	1.13
17.9	2×10^{-4}	8.0	1.21
20.1		7.2	1.07
13.9		37.4	1.10
17.9	2×10^{-3}	25.5	1.13
20.1		26.3	1.10
13.9	2×10^{-2}	124.6	0.18

Table III: Flow data from the literature.

Method	Alloy[†]	Temperature °C	$\dot{\varepsilon}$ sec^{-1}	Approximate Flow Stress MPa (@ ε)	Reference
Tension	80Z-20A	1450	2.78×10^{-4}	20 (0.95)	9
Tension	60Z-40A	1450	2.78×10^{-4}	30 (0.7)	9
Tension	40Z-60A	1450	2.78×10^{-4}	37 (0.43)	9
Tension	80Z-20A	1450	2.78×10^{-4}	19 (0.94)	4
Tension	80Z-20A	1450	2.78×10^{-3}	55 (0.49)	4
Constant Load	50v/oZ-50A	1500	2×10^{-3}[‡]	60 (--)	10
Compression	80Z-20A	1500	1.67×10^{-4}	16 (1)	11
Compression	80Z-20A	1500	1.67×10^{-3}	50 (0.2)	11
Tension	60Z-40A	1550	2.78×10^{-4}	10 (1.25)	9

†. Z represents ZrO_2 and A represents for Al_2O_3. Unless stated otherwise all alloys are in wt%.
‡. Creep rate under the applied stress.

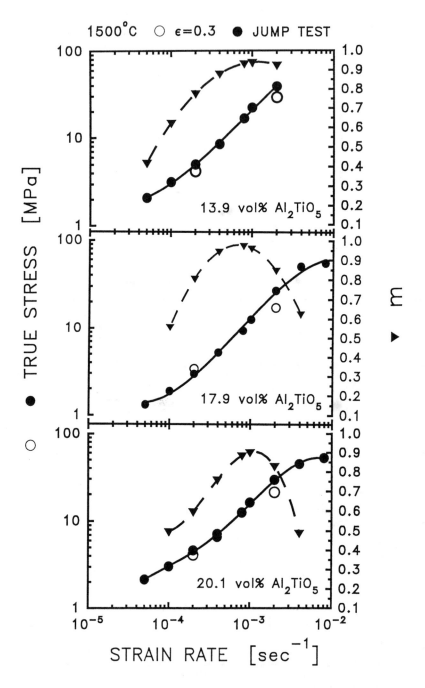

Figure 3. Strain rate sensitivity data for ceramic blends with varying Al₂TiO₅ contents.

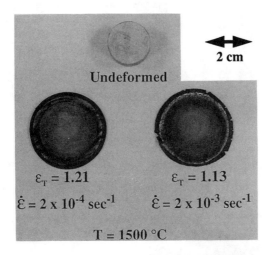

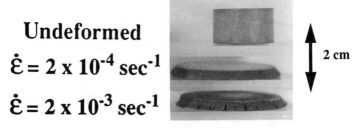

Figure 4. Geometry of typical samples in the undeformed and deformed states (two strain rates)

Two final points must be made. The first deals with the effectiveness of the procedure to maintain a constant strain rate while the second concerns the specimen geometry. It was observed that the molybdenum platens crept during testing. Nevertheless, the actual strains obtained in deformed specimens were generally within 0.05 of the desired strain. This error in the final specimen height can be shown to cause a 5% variation in the strain rate. Thus, the machine stiffness correction appears to be an adequate method of achieving true strain rates, at least over the strains encountered in this study.

It is recognized that with squat specimen geometries, frictional effects can require that the applied pressure necessary for material flow be much greater than the yield strength of the material. Thus, the effect of specimen geometry upon the observed flow stress must be addressed. Bishop [13] analyzed the effect of friction on the compression of circular cylinders. Assuming friction is constant across the faces of the cylinder, summing forces, and applying boundary conditions yields the analysis shown in Figure 6, where $\bar{p}$ is the average applied pressure, σ_0 is the yield strength, and a/h is the radius to height ratio.

As stated previously, the initial specimen geometry used in this study is very squat, with a/h≈0.89. During testing the a/h ratio can approach 4.55. Per Figure 6, depending on the coefficient of friction (μ), the applied pressure could approach three times the true flow stress of the material. Currently, μ is not known but some conclusions can be drawn from the edge

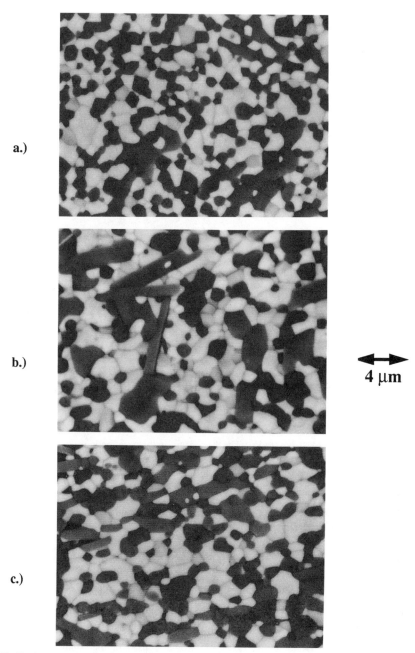

a.)

b.)

← → **4 μm**

c.)

Figure 5. Backscattered images showing the development of the microstructure in a 17.9 v/o Al_2TiO_5 blend a.) as-sintered, b.) deformed at 2×10^{-4} sec^{-1} ($\varepsilon = 1.21$), and c.) 2×10^{-3} sec^{-1} ($\varepsilon = 1.13$), respectively. ZrO_2 appears white, Al_2TiO_5 gray, and Al_2O_3 black.

geometry of the deformed specimens. Examination of Figure 4 shows that "barreling" is conspicuously absent, although the edges do have a "slope" associated with them. It is theorized that this sloping edge may be an artifact caused by a temperature gradient in the sample. Support for this theory is given by the observation that there is always an initial temperature drop in the molybdenum platen under the sample when the load is first applied through the top platen. Thus it seems likely that the top platen is somewhat colder than the bottom. This gradient could persist, at least during the initial stages of testing, causing different frictional constraints at opposite specimen surfaces. When samples were tested between graphite platens the edges had the classic bulge associated with medium to high frictional conditions. Thus the edge geometry in this case would lead one to conclude that the coefficient of friction is quite low, lending credence to the idea that the observed flow stress is approximately the true flow stress. The frictional coefficient will be measured in the near future using the ring upset procedure of Lee and Altan [15].

Table IV: Particle Size as determined by Quantitative Metallography

Volume Percent Al_2TiO_5[†]	ε	$\dot{\varepsilon}$ sec^{-1}	Particle Size		
			Al_2O_3 μm	ZrO_2 μm	Al_2TiO_5 μm
	Sintered	--	2.05	1.59	0.76
13.9	1.13	2×10^{-4}	2.04	1.93	1.49
	Sintered	--	2.18	1.16	1.13
17.9	1.21	2×10^{-4}	2.74	1.53	1.46
	1.13	2×10^{-3}	2.87	1.39	1.27
	Sintered	--	1.92	1.29	1.21
20.1	1.07	2×10^{-4}	2.20	1.55	2.62
	1.10	2×10^{-3}	2.07	1.21	1.60

†. Actual blends: 23.7 v/o Al_2O_3 - 62.4 % ZrO_2 - 13.9 % Al_2TiO_5
36.3 v/o Al_2O_3 - 45.8 % ZrO_2 - 17.9 % Al_2TiO_5
41.6 v/o Al_2O_3 - 38.4 % ZrO_2 - 20.1 % Al_2TiO_5

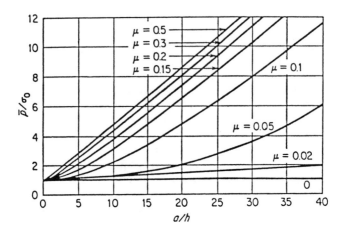

Figure 6. Average deformation pressure in compression of a disk as a function of μ and a/h [14].

<u>Summary and Conclusions</u>

This investigation has examined the effect of composition in Al_2O_3-ZrO_2-Al_2TiO_5 blends upon the flow behavior and microstructural coarsening at various strain rates. The initial experimental results suggest the following:

(1) The machine stiffness correction appears to be an adequate method to achieve constant strain rate compression testing.

(2) Although the specimens are very squat, frictional effects appear to be minimal. Thus, the flow stress observed is approximately the true flow stress.

(3) Ternary Al_2O_3-ZrO_2-Al_2TiO_5 blends have been shown to have low to moderate flow stresses at intermediate to high strain rates.

(4) The strain rate sensitivity for the alloys investigated is quite large (≥ 0.7) and appears to peak at relatively high strain rates (10^{-3} sec^{-1}).

(5) Edge cracking is observed at high strain rates, although it is minimal at low strains. It has been observed that cracked specimens can "heal" during deformation under certain conditions.

(6) The ternary microstructure is quite resistant to particle growth. Particle coarsening appears to be more sensitive to time at temperature than to the imposed strain rate.

<u>Acknowledgments</u>

The authors wish to acknowledge the generosity of Saila Karvinen of Kemira for kindly providing the titania powder used in this study and Caterpillar Inc. for financially supporting this research.

References

1. F. Wakai, S. Sakaguchi, and Y. Matsuno, "Superplasticity of Yttria-Stabilized Tetragonal Polycrystals,' Advanced Ceramic Materials, 1 (3) (1986), 259-63.

2. I-Wei Chen and Liang An Xue, "Development of Superplastic Structural Ceramics," Journal of the American Ceramics Society, 73 (9) (1990), 2585-2609.

3. T.G. Nieh, C.M. McNally, and J. Wadsworth, "Superplastic Properties of a fine-grained Yttria-Stabilized Tetragonal Polycrystal of Zirconia," Scripta Metallurgica, 22 1988), 1297-1300.

4. Fumihiro Wakai, "A Review of Superplasticity in ZrO_2-Toughened Ceramics," British Ceramic Transactions and Journal, 88 (6) (1989), 205-8.

5. Mr. B. McFall, private communication with author, Caterpillar Inc., September 1992.

6. Ervine E. Underwood, "Quantitative Metallography," Metals Handbook, Vol. 9, (The American Society for Metals, Metals Park, OH, 1985).

7. Mel I. Mendelson, "Average Grain Size in Polycrystalline Ceramics," Journal of the American Ceramics Society, 52 (8) (1969), 443-6.

8. T.G. Nieh and J. Wadsworth, "Superplastic Deformation Mechanisms in a Fine-Grained, Yttria-Stabilized Tetragonal Zirconia Polycrystal (Y-TZP)," Material Research Society Symposium Proceedings, Vol. 196, Editors M. Mayo, M Kobayashi, and J. Wadsworth, (Materials Research Society, Pittsburgh, PA, 1990), 331-6.

9. F. Wakai, Y. Kodama, S. Sakuguchi, N. Murayama, H. Kato, and T. Nagano, "Superplastic Deformation of ZrO_2/Al_2O_3 Duplex Composites," Proceedings of the MRS International Meeting on Advanced Materials, Vol. 7, Superplasticity, Executive Editors M. Doyama, S. Somiya, and R. Chang, (Materials Research Society, Pittsburgh, PA, 1989), 259-66.

10. B.J. Kellett and F.F. Lange, "Hot forging characteristics of transformation-toughened Al_2O_3/ZrO_2 composites," Journal of Materials Research, 3 (3) (1988), 545-51.

11. Fumihiro Wakai, Hidezumi Kato, Shuji Sakaguchi, and Norimitsu Murayama, "Compressive Deformation of Y_2O_3-Stabilized ZrO_2/Al_2O_3 Composite," Yogyo-Kyokai-Shi, 94 (9) (1986), 1017-20.

12. T. Epicier, G. Thomas, H. Wohlfromm, and J.S. Moya, "High resolution electron microscopy study of the cationic disorder in Al_2TiO_5," Journal of Materials Research, 6 (1) (1991), 138-45.

13. J.F.W. Bishop, "On the effect of Friction on Compression and Indentation between Flat Dies," Journal of the Mechanics and Physics of Solids, 6 (1958), 132-44.

14. George E. Dieter, Mechanical Metallurgy, Second Edition, (McGraw-Hill, New York, 1976), 564.

15. C.H. Lee and T. Altan, "Influence of Flow Stress and Friction Upon Metal Flow in Upset Forging of Rings and Cylinders," Journal of Engineering for Industry, ASME, Series B, 94 (1972), 775-82.

THE EFFECT OF MICROSTRUCTURAL EVOLUTION ON

SUPERPLASTICITY IN Ni$_3$Si(V,Mo)

Susan L. Stoner[+] and Amiya K. Mukherjee[*]

[+]Lawrence Livermore National Laboratory, University of California,
Livermore, California

[*]Department of Mechanical, Aeronautical and Materials Engineering,
University of California, Davis, California

Abstract

To further the understanding of superplasticity in intermetallics, this paper presents results of experimental investigations on an intermetallic alloy based on nickel silicide. Specifically, the evolution of the microstructure and its influence on superplastic performance is discussed. In the duplex microstructure, one phase showed grain growth, and the other, grain refinement. Cavitation occurred at interphase boundaries and final failure was by interlinkage of these cavities. Superplastic behavior was influenced by changing the orientation of the tensile axis. Though a transverse orientation showed more cavitation than longitudinal, it yielded greater elongation. An increased resistance to cavity coalescence in the transverse direction played a role in the enhanced ductility. Annealing the material improved the homogeneity of the microstructure. The annealed material showed improved strain-rate sensitivity values and enhanced superplasticity.

Advances in Superplasticity and Superplastic Forming
Edited by N. Chandra, H. Garmestani, R.E. Goforth
The Minerals, Metals & Materials Society, 1993

Introduction

Intermetallic alloys possess many attractive mechanical properties. Because of their ordered nature, dislocation motion is limited, particularly at elevated temperatures, resulting in exceptional high temperature strength. Intermetallics such as silicides and aluminides have the ability to form protective oxide films, and hence are highly resistant to oxidation and corrosion. Those intermetallics based on the lighter elements, like Ti_3Al, are low in density.

In spite of these attributes, interest in their use has been limited in the past because intermetallics show limited ductility. The inherent brittleness of these materials arises from a limited number of slip systems in their ordered crystal structure and a propensity for grain boundary embrittlement. Recently, progress has been made in improving the ductility of intermetallic alloys. Through control of the microstructure with alloying and thermomechanical processing, it has been shown that the ductility of several intermetallic systems can be substantially improved (1-5). With advances such as these, intermetallic alloys are promising candidates for use in advanced aerospace, turbine engine, and power plant designs.

Demonstration of superplasticity in intermetallics is of recent origin. Given their attributes, it is desirable to develop these materials for superplastic forming (SPF) applications. Microstructural features and mechanical properties must be well characterized to optmize the SPF process. An understanding of the effect of thermomechanical processing and test or forming parameters on mechanical behavior can be used to control superplastic performance.

This work considers an intermetallic alloy based on nickel silicide, $Ni_3Si(V,Mo)$. This alloy shows superplasticity at temperatures from 1273 - 1383 K (6) and at strain-rates from 10^{-4} to 1 s^{-1} (6,7). The focus is on the characteristics of the microstructure and its correlation to superplastic behavior.

Experimental Details

The composition of the material used for this study was (wt%) Ni-9Si-3.1V-4Mo. Vanadium and molybdenum were added to the binary Ni_3Si to promote superplasticity by stabilizing the β phase in the duplex microstructure.

A 25.4 mm x 25.4 mm x 127 mm ingot was produced by arc melting and drop casting of pure metal constituents. The thermomechanical processing of the material included a homogenization anneal, hot forging to a thickness of 15.3 mm, hot rolling to a thickness of 2.5 mm, and a final anneal at 1223 K for 16 hours. The final anneal was designed to restore maximum room temperature ductility.

Tensile specimens were cut from the material by electrical discharge machining. Specimens were prepared primarily with the tensile axis parallel to the rolling direction. (Unless otherwise stated, the results presented in this paper were obtained with specimens prepared in this orientation.) The final gage dimensions were 5.72 mm x 9.53 mm x 2.03 mm The specimens were tested using a digital controlled Instron 4505 tensile machine, interfaced with a data acquisition system. Tests were performed in an argon atmosphere surrounded by a radiant heat furnace.

The material was examined using optical metallography, scanning electron microscopy, and transmission electron microscopy. Cavitation was studied using a Ziess image analysis system.

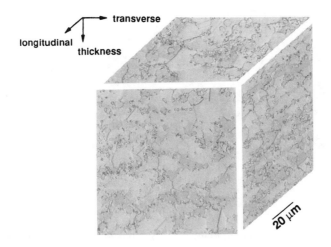

Figure 1. The as-received microstructure of Ni$_3$Si(V,Mo)

A Quantimet image analyzer was used to determine grain areas. A minimum of 250 grains were used for each measurement. The following equation given by ASTM E112, was used to convert grain area to a linear intercept value.

$$\text{Linear intercept} = .886(\text{grain area})^{0.5} \qquad (1)$$

Grain sizes are expressed as the linear intercept value measured in the plane normal to the sample thickness.

Results and Discussion

The microstructure of the Ni$_3$Si(V,Mo), as-received, is shown in Figure 1. The material has a duplex microstructure with a third phase present as precipitates. A large distribution of grain sizes and texturing in the direction of rolling is observed. Strings and large groups of grains of a single phase are apparent.

The dark grains are L1$_2$ cubic beta (β) phase. The average size of the β grains is 3.8 μm. The light grains consist of a phase mixture of alpha (α) - Ni solid solution phase and cubic β phase dispersions. This phase mixture will be referred to as ($\alpha+\beta$) from here on. The average size of the ($\alpha+\beta$) grains is 11.9 μm. The third phase is Mo-rich precipitates, 1 - 2 μm in diameter, which are randomly distributed throughout the microstructure. There is no apparent change in the size or volume fraction of these precipitates with deformation, suggesting they are insignificant to superplasticity in the Ni$_3$Si alloy.

Microstructural evolution

Strain-enhanced growth of the β grains and strain-enhanced refinement of the ($\alpha+\beta$) phase grains occurs during superplasticity. Hence, the two phases become more equal in size with strain. This behavior is demonstrated clearly in the micrographs in Figure 2.

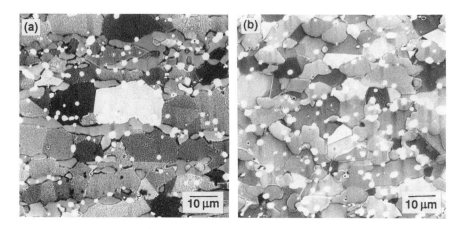

Figure 2. Microstructures (a) at the grip and (b) near the tip showing the longitudinal direction of a specimen tested at 1343K and $\dot{\varepsilon}$=10^{-3} s^{-1}. The tensile axis is horizontal.

The microstructure at the grip is shown compared to that near the tip (the direction of rolling is shown) for a specimen that was tested at 1343 K and a strain-rate ($\dot{\varepsilon}$) of 10^{-3} s^{-1}. These micrographs were made using back scattered electrons. The textured, multi-colored grains are the (α+β) phase. The single-colored grains are β phase and the white spots are the Mo-rich precipitates. It is also evident that grains that are initially elongated become more equiaxed with strain. These observations are consistent with the general tendency for the microstructure of superplastic materials to become more uniform with deformation.

True stress/true strain curves are shown in Figure 3 for constant strain-rate tests conducted at 1343 K and at strain-rates of: 6x10^{-4} s^{-1}, 1x10^{-3} s^{-1}, and 8x10^{-2} s^{-1}. Tensile stability is clearly demonstrated at the lower two strain-rates. (Flow stress and strain-rate sensitivity as a function of temperature and strain-rate have been presented elsewhere (8)).

The strain-hardening shown at the lower two strain-rates is related to dynamic grain growth in the β phase. Figure 4 shows β grain growth as a function of local strain (expressed as a reduction in area) for the three strain-rates discussed above. Dynamic grain growth was determined by subtracting the grain size in the grip from the grain size in the gage.

Growth of the β phase increases with decreasing strain-rate. The lack of grain growth at the highest strain-rate is consistent with the studies of Wilkinson and Caceras (9). In a number of superplastic materials they showed a tendency for the grain growth rate to reach a plateau at high strain-rates. This results because grain boundary migration is a time-dependent process and is therefore limited at high strain-rates.

Hamilton (10) has suggested that strain hardening is necessary for tensile stability and hence improves superplasticity. This is consistent with the results that have been presented. With decreasing strain-rate, both an increase in tensile stability and an increase in grain growth are shown.

Refinement in the ($\alpha+\beta$) phase was observed in all tests and appeared to be maximum at the lowest strain-rate. The mechanism of the refinement in the ($\alpha+\beta$) phase is uncertain at this time. No evidence of recrystallization was observed. Since the lowest strain-rate showed the best superplastic behavior, it appears that a maximum in grain refinement might be linked to a maximum in grain boundary sliding. Since long range cooperative grain boundary sliding brings about a redistribution of the phase structure, it might be that the ($\alpha+\beta$) phase is broken up in this process. Cooperative sliding has been discussed by several authors (11,12,13) and Yang, et.al., (13) attributed the breaking up of the α_2 phase in Ti$_3$Al alloys to this behavior.

Cavitation was observed in the Ni$_3$Si alloy during superplasticity. Cavities were located almost exclusively at β /($\alpha+\beta$) interfaces. For a given strain, an increase in cavitation with decreasing strain-rate was observed. The final specimen failure in all tests occurred by the interlinkage of cavities.

<u>Effect of test orientation and microstructural anisotropy</u>

It is clear from Figure 1 that grains are elongated in the direction of rolling in the as-received material. All of the results discussed thus far have considered testing with the tensile axis parallel to this direction. To assess the effect of microstructural anisotropy on the superplasticity of Ni$_3$Si(V,Mo), specimens were additionally prepared with their axes parallel to the transverse direction.

Figure 5 compares the stress/strain behaviors for loading in the longitudinal and transverse orientations. These tests were conducted at a constant strain-rate of 10^{-3} s^{-1} at 1343 K. The specimen tested in the transverse direction achieved greater terminal ductility while reaching a higher maximum stress.

Since the instability parameter is defined by the strain-rate sensitivity and the strain hardening coefficient, one or both of these values must be different for the two orientations. Strain-rate sensitivity (m) curves for the longitudinal and transverse orientations were generated. For strain-rates ranging from 10^{-4} to 10^{-1} s^{-1}, the m

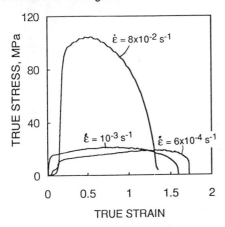

Figure 3. Stress/strain behavior for constant strain-rate tests conducted at 1343 K.

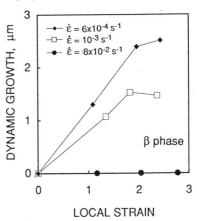

Figure 4. Strain-enhanced growth of the β phase for constant strain-rate tests conducted at 1343 K.

values for the two orientations were approximately the same. Thus, it appears that the enhanced ductility in the transverse direction cannot be attributed to a better strain-rate sensitivity. However, the total strain (ε) achieved in the tests used to develop the strain-rate sensitivity curves was low ($\varepsilon < 0.4$). The discrepancy may lie in the assumption that the calculated values of m are not sensitive to the structural details that vary with strain. At strains less than 0.4, where the strain-rate sensitivity curves were generated, the longitudinal and transverse curves in Figure 5 are similar. However, at a higher level of strain the two differ significantly. It might be that the strain-rate sensitivity in the longitudinal orientation falls off at a lower strain than it does in the transverse orientation. Alternately or additionally, the variation of the strain hardening coefficient with strain may differ for the two.

The growth of the β phase and refinement of the ($\alpha+\beta$) phase were similar in the two specimens. The cavitation characteristics for the two however were quite different. These characteristics might be different enough to influence the relative true strain-rate sensitivity values and/or the strain hardening coefficients. Numerous long cavity strings, primarily in the direction of the tensile axis, were present in the longitudinal specimen. Cavities nucleated at β /($\alpha+\beta$) interphases and tended to coalesce along the grain boundaries that were elongated in the direction of loading. Crack-like growth of cavities was observed as well.

The transverse specimen showed a greater resistance to cavity coalescence than the longitudinal specimen. Some cavity strings were observed but they were shorter and less numerous than in the longitudinal specimen. Crack-like growth was absent. Some coalescence of cavities into large holes was seen in the transverse specimen. Interestingly, the transverse orientation showed more cavitation, particularly at low and intermediate strains, than the longitudinal orientation. For specimens tested in both directions, the final failure occurred by cavity interlinkage. This is shown clearly in Figure 6 by the substantial cross sectional area and lack of extensive necking at the fracture tip of a specimen that achieved 387% elongation.

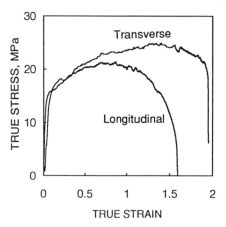

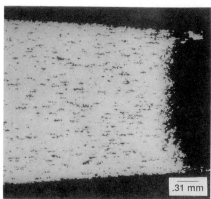

Figure 5. Stress/strain behaviors for the longitudinal and transverse orientations. Tests were conducted at $\dot{\varepsilon}$=10-3 s-1 and 1343 K.

Figure 6. Fracture tip of a specimen tested in the longitudinal orientation at $\dot{\varepsilon}$=10-3 s-1 and 1343 K.

From the observations discussed above, the following explanation of the relative behaviors shown in Figure 5 is proposed. At lower strains, the stress/strain behavior of the two orientations are similar as reflected by their comparable strain-rate sensitivities. At higher strains, the influence of the cavitation characteristics on the instability parameters becomes significant. At a true strain of approximately 0.8, the coalescence of cavities becomes sufficient in the longitudinal orientation to initiate necking and the curve begins to drop off. Because of a greater resistance to cavity interlinkage in the transverse orientation, the strain hardening behavior continues and initiation of necking is delayed to a higher strain. This results in a higher maximum flow stress and a larger terminal ductility in the transverse direction.

<u>Effect of annealing</u>

As discussed earlier, the as-received $Ni_3Si(V,Mo)$ microstructure shows considerable anisotropy. It is well established that ideal superplastic microstructures contain fine, uniform grains. This suggests that improvement in the initial microstructure of the Ni_3Si alloy through thermomechanical processing might lead to improvement in superplastic performance.

The as-received material was annealed for four hours at 1323K. (From here on "annealed" will refer to this particular anneal.) The grain size distributions of the β and $(\alpha+\beta)$ phases resulting from the anneal are compared to those in the as-received microstructure in Figure 7. The distributions of both phases in the annealed condition

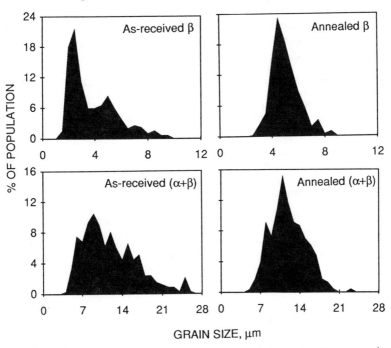

Figure 7. Grain size distributions of the β and $(\alpha+\beta)$ phases for the as-received and annealed microstructures. Grain sizes are expressed as a linear intercept value measured in the plane normal to the specimen thickness.

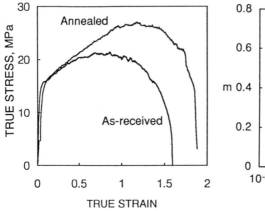

Figure 8. Stress/strain behaviors for the as-received and annealed conditions. Tests were conducted at $\dot{\varepsilon}=10^{-3}$ s^{-1} and 1343 K.

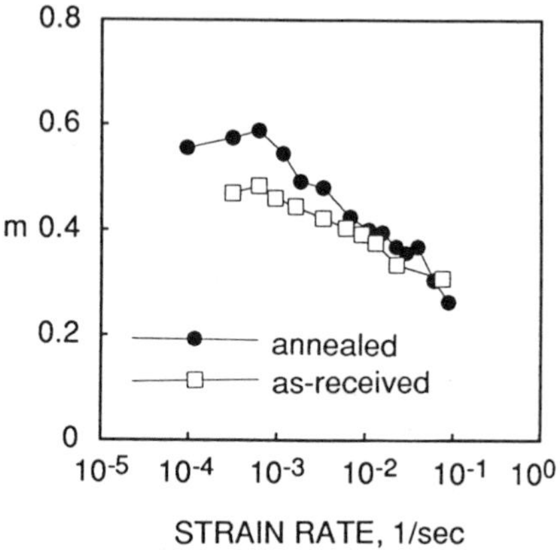

Figure 9. Strain-rate sensitivites at 1343 K for the as-received and annealed microstructures.

are narrower than those in the as-received condition. In the β phase, a shift towards larger sizes is apparent, owing to static grain growth. The resulting average size of the β grains in the annealed material is 5.0 μm. That of the $(\alpha+\beta)$ grains remained at 11.9 μm. This latter result indicates that deformation is necessary to induce $(\alpha+\beta)$ refinement in Ni$_3$Si(V,Mo).

The grains in both phases are more equiaxed in the annealed material. Additionally, the two phases are more randomly distributed following the anneal, i.e., there are not as many strings or groups of grains of a single phase. Mechanical twins which are present in the as-received microstructure are not as numerous in the annealed microstructure.

The stress/strain behavior of the Ni$_3$Si alloy is compared for the as-received and the annealed conditions in Figure 8. More strain was achieved in the annealed material. This is an expected result since the discussed changes in the microstructure due to the anneal tend to be favorable for superplasticity.

Figure 9 compares the strain-rate sensitivities for the as-received and annealed conditions. The strain-rate sensitivity for the annealed material was higher than that of the as-received material over a wide range of strain-rates. At the highest rates, where the m values are equal, deformation is no longer governed by superplasticity. Assuming the strain-rate sensitivities for the two conditions show the same trend with increasing strain, these curves are consistent with the stress/strain behaviors shown in Figure 8.

A greater amount of strain hardening is observed in the annealed material. The amount of strain-enhanced growth of the β phase grains in the as-received and annealed materials was approximately the same. However, because the initial size of the β grains was larger in the annealed material, this may at least partially explain why it shows more hardening.

In addition to grain size and uniformity of the phases, it is likely that there are other factors contributing to the enhanced ductility observed in the annealed material. One such factor might be the structure of the grain boundaries. It is known that the mechanical behavior of metals and alloys is influenced by grain boundary structure as well as grain boundary area.

In the intermetallic TiAl, Imayev and coworkers (14) showed that superplastic performance was deteriorated by the presence of twin boundaries. In contrast to "random" boundaries, grain boundary sliding is impeded by twin boundaries which trap lattice dislocations, even at high temperatures. Using thermomechanical processing, Imayev, et.al., varied the proportionate number of random boundaries to twin boundaries in TiAl samples. For those with a higher percentage of the random type, better strain-rate sensitivity values and greater superplastic elongations were shown. For the present case, it might be that annealing the $Ni_3Si(V,Mo)$ reduced the number of twin or other types of "special" boundaries, which impede grain boundary sliding, in the microstructure. The annealed microstructure could then have a larger percentage of "random" boundaries than the as-received microstructure, which would better facilitate superplasticity.

The cavitation was measured in the tested as-received and annealed specimens. At higher levels of strain, the two showed approximately the same amount of cavitation. At low strains however, the annealed specimen showed less cavitation, owing most likely to the more uniform starting structure.

Conclusions

Results of experimental investigations on the microstructural evolution in superplastic $Ni_3Si(V,Mo)$ have been presented. The microstructure plays a significant role in the superplastic behavior of this material.

At the lower strain-rates tested, strain-enhanced growth occurred in the β phase grains during superplastic deformation. The stress/strain curves show strain hardening as a result of grain growth. At the highest strain-rate, β grain growth was absent and the stress/strain curve showed no tensile stability. Refinement was observed in the $(\alpha+\beta)$ grains at all strain-rates tested.

Test orientation and hence microstructural anisotropy had a significant effect on the superplastic behavior of this alloy. For tension in the transverse direction, more total strain was achieved and more strain hardening was observed than for tension in the longitudinal direction. The superior tensile stability demonstrated in the transverse orientation is attributed to a greater resistance to cavity interlinkage.

Annealing the material resulted in a more uniform starting microstructure. The annealed microstructure showed improved strain-rate sensitivity values and a greater total elongation than the as-received microstructure. The structure of the grain boundaries might play a role in the enhanced superplasticity seen in the annealed microstructure.

References

1. K. H. Hahn and K. Vedula, Scripta Metall., 23, (1989),7.

2. C.T. Liu, C.L. White, C.C. Koch, and E.H. Lee, in High Temperature Materials Chemistry II, Vol. 83-7, edited by Munir Cubicciotti , The Electrochem. Soc. Inc., (1983), 32.

3. K. Aoki and O. Izumi, Nippon Kinzoku Gakkaishi, 43, (1979), 1190.

4. M.J. Blackburn and M.P. Smith, Air Force Technical Report, AFWAL-TR-81-4046 (1981).

5. J. Wadsworth and F.H. Froes, J. Metals, 41, (1989), 12.

6. T.G. Nieh, in Superplasticity in Metals, Ceramics, and Intermetallics, edited by M.J. Mayo, M. Kobayashi and J. Wadsworth, MRS, Pittsburgh, PA, (1990), 189.

7. S.L. Stoner, "Superplasticity in a Nickel Silicide Alloy, $Ni_3Si(V,Mo)$" (Masters thesis, University of California at Davis, 1984).

8. S.L. Stoner and A.K. Mukherjee, Materials Science and Engineering, A153 , (1992), 465.

9. D.S. Wilkinson and C.H. Caceres, Acta Metall., 32, (1984),415.

10. C.H. Hamilton, in Strength of Metals and Alloys, edited by H.J. McQueen, J.P. Bailon, and J.I. Dickson, Pergamon, Oxford, (1986), 831.

11. M.G. Zelin and M.V. Alexsandrova, in Superplasticity in Advanced Materials, edited by S. Hori, M. Tokizane, and N. Furushiro, Japan Society for Research on Superplasticity, Osaka Japan, (1991), 63.

12. R.Z. Valiev and M.G. Zelin, in Superplasticity in Advanced Materials, edited by S. Hori, M. Tokizane, and N. Furushiro, Japan Society for Research on Superplasticity, Osaka Japan, (1991), 95.

13. H.S. Yang, M.G. Zelin, R.Z. Valiev, and A.K. Mukherjee, Scripta Metall., 26, (1992), 1707.

14. R.M. Imayev, O.A. Kaibyshev, and G.A. Salishchev, Acta Metall., 40, (1992), 581.

Acknowledgments

The authors would like to express their appreciation to Warren Oliver of Oak Ridge National Laboratory, who provided the material for this study. We would like to thank Scott Preuss, of Lawrence Livermore National Laboratory, for his efforts in running the tests. We are also grateful to Kerry Cadwell and Rick Gross for their assistance in making the grain size measurements. This work was performed under the auspices of the U.S. Department of Energy by Lawrence Livermore National Laboratory under contract No. W-7405-Eng-48.

MICROSTRUCTURES AND SUPERPLASTICITY IN NEAR-GAMMA TITANIUM

ALUMINIDE ALLOYS

C.C. Bampton and P.L. Martin

Rockwell International, Science Center
1049 Camino Dos Rios
Thousand Oaks, CA 91360

Abstract

Microstructure control by thermomechanical processing in near-gamma titanium aluminide alloys has recently progressed to a point where we are able to reliably produce a wide range of microstructures in a single alloy. We are now studying the basic superplastic deformation behaviors of Ti-48Al-2Cr-2Nb-1(Ta or Mo) (at%) alloys, processed to widely different microstructures. Correlations are made between microstructural details and flow stress, strain hardening, strain-rate hardening, necking, cavitation and failure. Special emphasis is given to the cavitation behavior since this phenomenon may constitute a major limitation to the useful application of superplastic forming for gamma TiAl structures.

Advances in Superplasticity and Superplastic Forming
Edited by N. Chandra, H. Garmestani, R.E. Goforth
The Minerals, Metals & Materials Society, 1993

Introduction

Near-γ titanium aluminide (TiAl) intermetallic alloys are being actively pursued as low density replacements for superalloys in structures operating in the useful temperature range 700 - 900°C (Ref.1). It is clear that these alloys, to be successful, must be regarded as engineered materials in the sense that sophisticated thermomechanical processing is necessary both to provide the desired shapes and the engineering properties. The alloys are based on the ordered $L1_0$ γ-TiAl phase but all the alloys of interest contain small volume fractions of the ordered α_2-Ti3Al phase and in some cases an ordered β phase. The presence of even small volume fractions of these second and third phases are critical for enabling microstructural refinement and modification of mechanical behavior. This is accomplished by using our knowledge of phase transformation temperatures and kinetics during thermomechanical processing. Solute segregation (most importantly Al segregation) is inevitable in the cast ingot, due to peritectic solidification reactions. It is not practical to remove this segregation by static thermal treatments due to the slow diffusion rate of Al in γ and the tendency for grain growth at very high temperatures. The segregation can persist through conventional, single-step, thermomechanical processing. The segregation results in profound modifications of local phase stabilities and microstructures due to the sensitivity of transus temperatures to Al concentration.

While activities are ongoing in this laboratory to study the processing/microstructure/properties relationships in ductile, near-γ alloys (Ref. 2) we are concurrently studying their superplastic deformation behaviors. Superplastic deformation is recognized as a potentially useful means of economically shaping difficult-to-process materials such as γ-TiAl. It is imperative to understand and exploit or control the opportunities for microstructural evolution during thermomechanical processing both prior to superplastic forming (SPF) and during SPF itself.

This is an interim status report on some of our work-in-progress. It is our intent to publish a more extensive report and analysis in the future covering a wider range of compositions and microstructures.

Experimental Procedure

Vacuum arc remelting produced an approximately 10 kg ingot of the composition shown in Table I.

Table I. Chemical Compositions Measured from Ingot in Atomic Percent

Location	Al	Nb	Cr	Ta	C	O	N	H
Top	47.05	1.98	1.98	0.95	.037	.158	.021	.120
Bottom	44.16	1.84	1.98	1.24	.040	.170	.017	.135

A right cylinder of the cast ingot was isothermally upset forged 73% at 1150°C and a (constant) strain rate of 1.6×10^{-3} sec^{-1}. The height after this forging step was approximately 45 mm.

Wire EDM was used to remove 1.8 mm thick slices from the forged pancake with thin dimensions perpendicular to the forging axis so that the "width" of the "sheet" samples was ~ 45 mm.

The "sheet" samples were heat treated at 1290°C/8 hours to recrystallize (in quartz backfilled with Ar and Ta wrapped). The samples were ground to a final thickness of ~1 mm.

Microstructural characterization utilized standard optical metallography and analytical electron microscopy. Backscattered electron images (BEI) were used for the bulk of the study since they convey both chemical and morphological information. TEM specimens were made by twin-jet electropolishing and foils examined on a Philips CM30 at 300 KV.

Sub-scale "dog-bone" test coupons were wire-EDM machined from the "sheet" slices as indicated schematically in figure 1. These coupons were suitable for our custom-built superplastic tensile test system which has retort systems capable of providing vacuum or inert environments pressurized to a maximum of 600 psi. Vacuum was used for all the tests reported here.

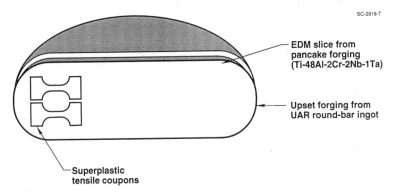

SC-2918-T

EDM slice from pancake forging (Ti-48Al-2Cr-2Nb-1Ta)

Upset forging from UAR round-bar ingot

Superplastic tensile coupons

Figure 1. Schematic illustration of superplastic coupon orientation.

Step-strain rate tests were carried out at each of the specified test temperatures. This test is computer-controlled to sequentially determine the initial plastic flow stress at each of a range of constant, true strain rates. The load and extension data are collected by the control computer and processed to give the following: true stress versus true strain rate data, a least-squares-fit, third or fourth order polynomial equation for log (true stress) versus log (true strain rate) and strain rate sensitivity index (m-value) versus log (true strain rate).

Constant true strain rate tests were carried out at each of the specified test temperatures with strain rates selected by analysis of the step strain rate test data. This test is computer-controlled to impose a constant true strain rate on the test coupon over the complete strain range to failure of the coupon. The load and extension data are collected by the control computer and processed to give a graphical plot of true stress versus true strain.

Figure 2 illustrates the pre-SPF processing temperatures and the three SPF testing temperatures in relation to the boundaries in the binary Ti-Al phase diagram.

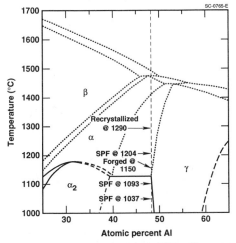

Figure 2. Pre-SPF processing temperatures and three SPF testing temperatures in relation to the boundaries in the binary Ti-Al equilibrium phase diagram.

99

Results

Figure 3 shows an example (at 1037°C) of the results from a step-strain rate test and figure 4 summarizes flow stress variations with temperature and strain rate.

SC-2913-T

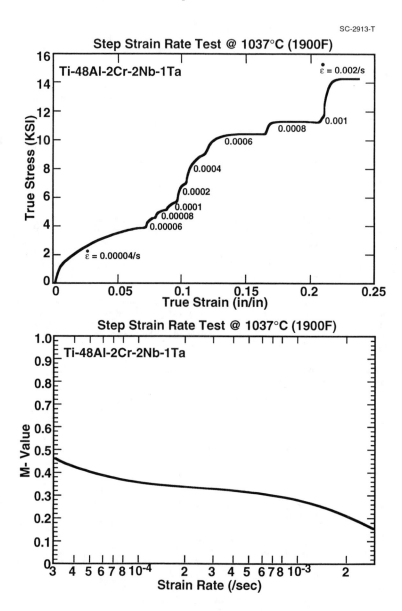

Figure 3. Strain rate sensitivity in Ti-48Al-2Cr-2Nb-1Ta (at.%) @ 1037°C

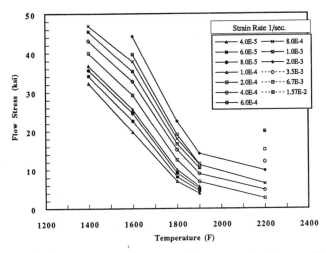

Figure 4. Flow stress versus temperature for a range of strain rates with the
Ti-48Al-2Cr-2Nb-1Ta (at.%) alloy

Table 2 summarizes the results of tensile tests at true constant strain rate of 0.0002/sec at three different temperatures. All three tests showed limited ductility (relative to other SPF materials), with diffuse necking from the shoulders to the fracture surfaces of the tensile coupons.

Table II. Superplastic Tensile Tests to Failure at Constant True Strain Rate = 0.0002/sec.

Temperature (°C)	Elongation (%)	Reduction in Area (%)
1037	99	63
1093	73	66
1204	89	58

Figure 5 illustrates the main features of the microstructure evolution with superplastic deformation at 1037°C and 0.0002/sec. The forged microstructure is dramatically broken-up with increasing superplastic strain, although the segregation, evidenced by the light, Ti-rich regions and dark, Al-rich regions, persists all the way to fracture. It is also clear that cavitation initiates at very low strains, principally in the Ti-rich regions, and grows with strain until cavity interlinkage causes final rupture of the coupon.

Figure 6 clearly shows the initial segregation pattern in the forging (a), and the profound modification of the microstructure by the superplastic deformation (b versus c). It should be noted that figs. 6 a, b, and c represent identical thermal histories, the sole difference being the superplastic strain. The micrograph in fig. 6b clearly shows the three phases (identified by TEM): the dark phase is equiaxed, single phase γ; the very light, generally inter-granular particles, are single phase β; and the lamellar colonies are alternating platelets of γ and α_2. Fig. 6c, appears to show complete break-up of the lamellar colonies, a redistribution of the β phase particles and dynamically recrystallized γ grains (or sub-grains).

Figure 7 shows a typical example of a very early stage of cavitation (from the same coupon shown in figs 5 and 6). The cavity appears to be formed by a debonding process along an interface of a lamellar colony and a γ grain (or intergranular β particles) which is oriented perpendicular to the applied tensile stress. Figure 7b shows the final fracture surface. The scale of the predominant features in the fracture surface are similar to the scale of the grain structure seen in figure 6c (recrystallized grains or subgrains of single phase γ).

101

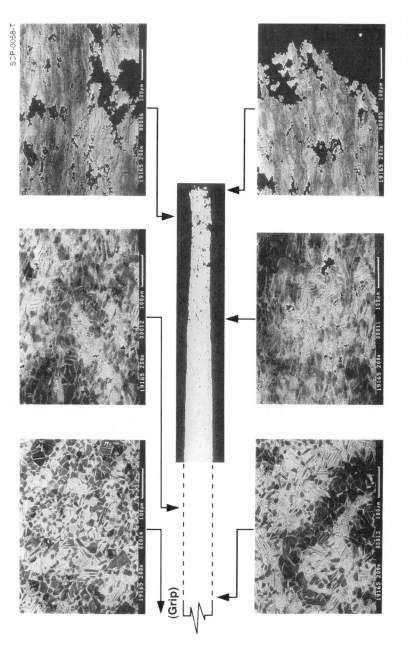

Figure 5. Cross-section back-scatter scanning electron micrographs of Ti-48Al-2Cr-2Nb-1Ta tensile coupon after failure @ 1037°C and 0.0002/sec.

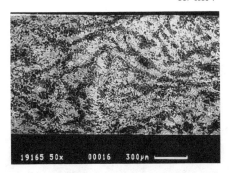

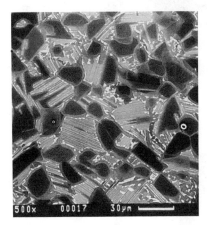
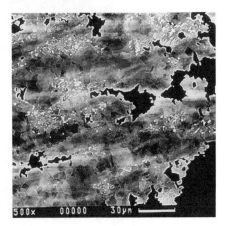

Figure 6. Back-scatter scanning electron micrographs of Ti-48Al-2Cr-2Nb-1Ta tensile coupon after failure @ 1037°C and 0.0002/sec (a) showing segregation patterns in the coupon grip; (b) and (c) showing microstructural changes from grip to fracture respectively

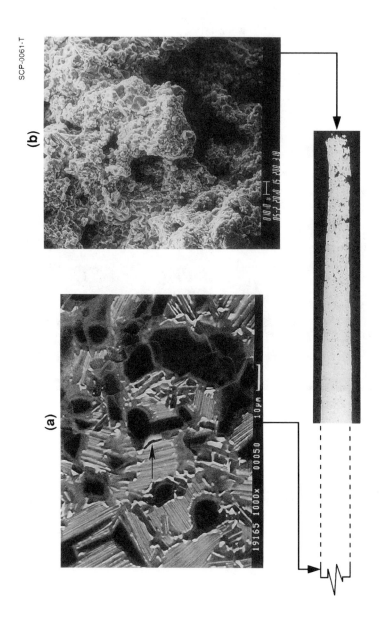

(b)

(a)

Figure 7. Backscatter SEM showing early cavitation (a) and secondary electron SEM showing final fracture surface (b) in the superplastic tensile coupon shown in fig. 5

104

Superplastic tensile tests at constant true strain rate of 0.0002/s at 55°C higher temperature (1093°C) produced essentially the same microstructural evolution as the 1037°C tests discussed above. A further test temperature rise to 1204°C, however, caused significantly different microstructures. Figure 8 shows this case for direct comparison with the 1037°C case shown earlier in fig. 5. The main effect is the transformation of the majority of the lamellar colonies to single phase grains in the statically annealed grip region of the coupon. Some overall static grain growth (of the γ grains) is also apparent with a significant amount of annealing twins.

Discussion and Conclusions

It was anticipated that the testing at 1093°C would show some significant differences relative to 1037°C since, as indicated in fig. 2, the alloy would have a larger volume fraction of γ. At the lower temperature it should still have sufficient second phase, α_2 to inhibit grain growth. Similar behavior was found, however, at both temperatures. The major effects were the complete break-up of the lamellar colonies and the progressive growth of cavities. The cavities appeared to be predominantly nucleated at lamellar colony boundaries oriented normal to the applied tensile axis.

Further inspection of the phase diagram in fig. 2, suggests that the third (and highest) SPF testing temperature (1204°C) may have a significantly higher volume fraction of second phase (disordered α in this case) than at each of the lower test temperatures. However, the alloy appeared to undergo significant static recrystallization at the 1204°C temperature which resulted in elimination of the lamellar colonies, formation of significant numbers of twin boundaries and a general coarsening of the microstructure.

A general conclusion from these results is that the alloy thermomechanically processed to the duplex microstructure (a mixture of equiaxed γ and lamellar colonies) is not suitable for SPF since the microstructure is not stable against dynamic transformations below the eutectoid temperature nor against static transformations above the eutectoid. Furthermore, the generally high flow stresses below the eutectoid and the presence of the lamellar colonies, encouraged cavity formation very early in the superplastic deformation process with rupture occurring at low strains by cavity linkage.

This kind of information is valuable for a more systematic approach to development of high temperature thermomechanical processing of the γ-TiAl alloys, including forging, extrusion, rolling and consolidation of TiAl-based fiber-reinforced composites, as well as the SPF process discussed here.

References

1. Y. Kim, J.O.M., 43, (1991), pp 40-47.

2. P.L. Martin and C.G. Rhodes, "Microstructure and Properties of Refractory Metal Modified Ti-48at%Al-2at%Nb-2at%Cr", to be published in Proc., 7th World Conf. on Titanium. Eds., I.L. Caplan and F.H. Froes, TMS-AIME, San Diego, CA, 1993.

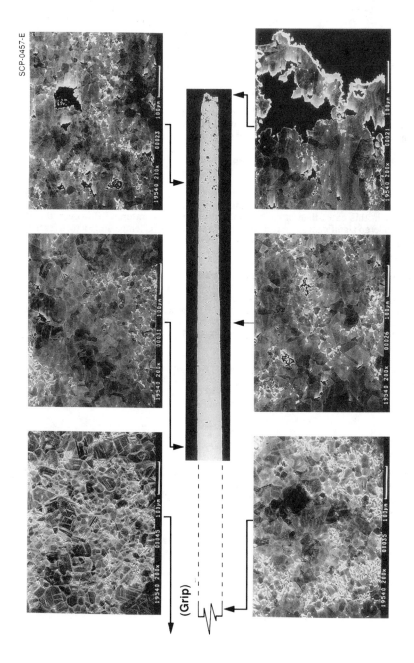

SCP-0457-E

(Grip)

Figure 8. Cross-section back-scatter scanning electron micrographs of Ti-48Al-2Cr-2Nb-1Ta tensile coupon after failure @ 1204°C and 0.0002/sec.

NOVEL EXPERIMENTAL TECHNIQUES
AND PROCESS MODELING

Investigation of Deformation Mechanisms of Superplastic Ti-6Al-4V using Back Scattered Kikuchi Diffraction Patterns.

H. Garmestani, G.S. Sohi, S. Mukherjee, N. Chandra

Department of Mechanical Engineering

FAMU/FSU College of Engineering

Tallahassee, Fl 32316

ABSTRACT

In this paper a method to investigate the contributions of different deformation mechanisms towards total strain is proposed. The components of strain due to diffusion dominated grain boundary sliding and dislocation dominated plastic deformation are resolved using Back Scattered Kikuchi diffraction patterns. Using this technique, pole figures in superplastic materials stretched to different strain levels are evaluated. From this data, orientation distribution functions and frequency misorientation plots are determined, which are then used to predict the contributions of different deformation mechanisms towards total strain.

1. Introduction

The impetus to reduce the manufacturing costs of complex components has resulted in considerable attention being devoted in recent years to the phenomenon of superplasticity and superplastic forming. In this process elongations as high as 8000% have been recorded. The ideal microstructure for superplastic deformation consists of small (10 micron) equiaxed grains [1]. During superplastic deformation a gradual increase in grain size can occur and any initial alignment of the microstructure will disappear[2]. The findings in texture studies of superplastically deformed materials resulted in very ambiguous interpretations of the operating mechanisms[3]. It is generally agreed that the major mode of superplastic deformation is grain boundary sliding with diffusional accomodation mechanisms acting to maintain grain compatibility; also dislocational slip and creep contribute to overall strain in some ranges of superplastic deformations. It is still an open issue which of these mechanisms are active in the contribution towards the overall strain. This issue has presently assumed greater importance in order to explain superplasticity in a variety of exotic systems including metal matrix, intermetallic matrix materials, ceramics both in monolithic and composite forms. Also the observance of superplasticity in high strain rates (10^{-3} to 10^{2} per sec) has brought this fundamental question to the foreground [4].

In this work a superplastic texture analysis is performed on Ti-6Al-4V using the back scattered Kikuchi technique. Ti-6Al-4V has been increasingly used in steam

Advances in Superplasticity and Superplastic Forming
Edited by N. Chandra, H. Garmestani, R.E. Goforth
The Minerals, Metals & Materials Society, 1993

turbines, aircraft gas turbines and aircraft body structures. Ti-6Al-4V is an $\alpha + \beta$ alloy with vanadium acting as a β stabilizer (BCC) and aluminum as an α stabilizer (HCP). During superplastic deformation at a temperature of 927°C this material exists in its $50\% - 50\%$ volume ratio for α and β phases. However, at room temperature this proportion will change depending on the cooling rate and the processing condition.(in favor of 90% α+10%β).

Texture analysis of the α phase in Ti-6Al-4V can be used to study the deformation mechanisms present in the superplastic ranges.The texture formation and development for Ti-Al alloys are slightly different than those of most other low c/a HCP materials which have their basal (0001) planes concentrated in a region 30-40° from the normal in the transverse direction [5]; Ti-6Al-4V has a cold rolling texture which has the basal poles aligned in the normal direction (basal texture) [6,7]. In general, the deformation mechanism in HCP metals are not as symmetrically distributed as in cubic metals. The classic textures observed in the Ti-Al system are split TD and basal textures. Careful attention should be given with regards to the texture development since as a result of appearance and disappearance of split rolling direction texture at different stages of straining may be mistaken with other deformation mechanisms [8].

A strong deformation texture is created as a result of the low number of slip systems and a large amount of deformation twinning. The interactions between deformation mechanisms and texture formation on the one side and deformation mechanisms and mechanical anisotropy on the other are different for materials forming under superplastic conditions and other deformation processes. In general most HCP materials with low c/a ratios, slip is activated on prismatic and pyramidal planes in a direction $[11\bar{2}0]$. At higher temperatures basal slip on $\{10\bar{1}1\}$ is also observed. The complicated interaction between slip and twinning is the cause for a marked deformation texture. Under tensile stresses in c-direction, primarily $\{10\bar{1}2\}$ twins and sometimes $\{11\bar{2}2\}$ twins are activated and under compressive loading in a c-direction, $\{11\bar{2}2\}$ twinning and $\{10\bar{1}1\}$ twinning at elevated temperatures are observed [9].

In general, grain boundary sliding is believed to cause a loss of texture in superplastic materials [2]. There is also an evidence of slip due to rotation about certain directions [11]. A study of the evolution of the texture components in the two phases of α and β with 50%-50% ratio becomes important to investigate the mechanisms of deformation responsible for superplastic forming . Due to the lack of possible slip systems in the α phase the contribution from the strains in the β phase (bcc) alters the expected texture [9,11]. While many investigators have reported on different deformation mechanisms of Ti-Al alloys, no information is available for the case in which there is a high ratio of the β phase present in the matrix. It is reported however that there is a texture loss in both phases and that the loss due to the β -phase being greater than that of α -phase [1]. In these studies however the nature of such texture

110

changes are not reported. For the case of Ti-6Al-4V as a result of the phase transformation from a 50-50% for α and β phases at the superplastic temperatures (927°C) to 90% α and 10% β, about 45% of the measured texture of the α-phase is due to the superplastic deformation of the β-phase [1]. Also it should be remembered that in this two phase alloy α-phase is considered to be the hard phase and the corresponding texture change will then be due to the 45% of β-phase in the form of α-phase.

A detailed study of superplastic deformation of Zn-40% Al has been performed by Melton et.al [12]. This study is important since this is the only detailed study of a hexagonal system available. The Zn rich phase shows a strong $(\bar{1}013)$ [2111] and $(1010)[0001]$ fiber texture and also $\{\bar{1}010\}$ $\langle 1\bar{2}10 \rangle$ orientation. In that study it was shown that in general texture was decreased as the material is deformed at superplastic rates. This study shows that the $\langle 1\bar{2}11 \rangle$ fiber texture is almost removed after deformation at all strain rates, but the [0001] fiber texture was strengthened after deformation. Also it was shown that $\{\bar{1}010\}$ $\langle 1\bar{2}10 \rangle$ was reduced at all strain rates but more rapidly at the lowest strain rates. This component of texture and the [0001] fiber texture did not seem to change as a result of annealing but the $\langle 1\bar{2}11 \rangle$ fiber texture was considerably reduced.

2. Texture Measurements Using BKD

Texture defines the volume fraction of crystals in a certain orientation independent of their spatial arrangement. Texture is one of the fundamental parameters characterizing polycrystalline materials in addition to crystal structure and lattice defects. This also applies to superplastic materials which require very fine grains (5 microns average grain size). Evolution of anisotropy and rotation of crystal axis during superplastic deformation are not well characterized. Micro texture studies provide an excellent and convenient means to study these processes themselves or to obtain information about the history of deformation in the superplastic material.

The Orientation Distribution Function (ODF), expresses the probability that a crystallite has an orientation (ϕ_1, ψ, ϕ_2) with respect to the specimen axes. The availability of ODF's with a high accuracy allows the calculation of orientation mean values of physio-chemical properties in the same range of accuracy, e.g. the Young's modulus, Yield locus, magnetic anisotropy or others. In this paper the Roe's notation is adopted [13]. Some of these properties can be evaluated using the single crystal property with orientation distribution function acting as the weight function in the mean value formula. However properties that are influenced by grain interactions are not taken into account by the simple orientation mean value. In this case the mutual arrangement of crystals of various orientations is also important in addition to the classified texture functions. Also a large number of "generalized texture quantities" should also be included in consideration such as inhomogeneous texture, correlation functions, boundary orientations misorientation angle, and others. These generalized textural quantities are similar to the texture itself in as far as they contain orientational parameters but additionally they also contain spatial parameters. Investigation of the evolution of such parameters in addition to ODF during superplastic forming will enhance our fundamental understanding of the process itself. Experimental texture investigations in metals have been carried out since the discovery of x-ray diffraction and even earlier. Texture can be studied qualitatively using pole figures. Quantitative texture studies involve the use of orientation distribution function (ODF). The theoretical methods of texture analysis originated from the problem of pole figure inversion, the calculation of the orientation distribution function from experimentally available pole figures [14]. In order for pole figure inversion to be unique, complete pole figures are needed. This is not possible since it is only desirable to obtain incomplete pole figures for experimental reasons [15]. The problem of pole figure inversion using incomplete pole figures is usually associated with the ghost phenomena. Ghost phenomena in the ODF are due to the superposition of pole figures related to each other by crystal symmetry operations[16,17]. This problem along with the problem of the pole figure normalization diverted the attention of many to measure ODF's directly.

Convergent beam electron diffraction patterns in Transmission Electron Microscopy (TEM) and Backscatter Kikuchi Diffraction (BKD) to obtain Electron Backscatter Diffraction Patterns (EBSP) inside a Scanning Electron Microscope (SEM) produce very valuable information about the crystal structure of the individual crystalline grains. Using such techniques it is possible to determine Orientation Distribution Functions directly from the Kikuchi line and pole information from the polycrystalline materials. Pole figures for the microstructure, relative grain boundary misorientation, tilt and twist can also be obtained from this method.

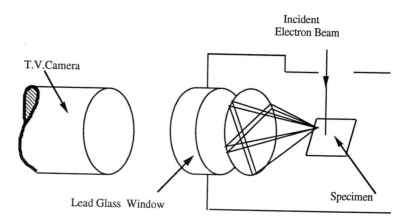

Figure 1: Schematics of BKD Technique

Backscatter Kikuchi Diffraction Technique

Dingley [5] developed an on line computer assisted technique for electron backscatter, or backscatter Kikuchi diffraction (BKD) analysis following the initial work of Venables [7]. In the spot mode of an ordinary scanning electron microscope, a diffraction pattern originates by inelastic scattering in a volume near the surface of the specimen. The dimension of this volume may be as small as 100-200 nm, depending upon the materials used. Electron backscatter diffraction differs from other diffraction techniques used in the scanning electron microscope in that the patterns are generated using a stationary probe. In Dingley's method, this image is projected onto a phosphor screen located inside the vacuum chamber and also through an optical port onto a high-gain television camera (ISIT) which is interfaced to a computer. Figure 1 is a schematic diagram illustrating the arrangement adopted in the SEM for imaging the patterns, and Figure 2 is a typical pattern from the α-phase of the Ti-6Al-4V as viewed on the imaging phosphor screen.

The analysis for orientation is accomplished by identifying some common crystallographic zone axes in the pattern; two axes are required in cubic materials, for example. With these axes identified, the computer can easily calculate the lattice orientation of the probed crystallite.

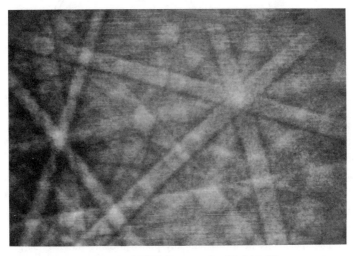

Figure 2: Typical BKD Pattern for Ti-6Al-4V

Misorientation Measurement

One of the microstructural parameters to consider in such studies is the distribution of grain boundary misorientation. The texture of a material, as described by pole figures or by ODF only provides information on the frequency, with which certain orientations are present in the material. These parameters contain no information about the "orientation topology", i.e. the spatial arrangement, the size and the form of grains with specific orientation. Therefore, it is impossible to calculate the orientation correlation between neighboring grains from the texture. It is obvious that there are strong links between texture and orientation correlation. For example, in a two phase material, in which each phase has one texture, one will expect that the average orientation relation between the two phases is determined by the relationship between the two textures. Recent studies on superplastic deformation of materials revealed that the misorientation associated with the grain boundaries are high angle. In general it is found that for superplasticity grain boundaries with low tilt and high twists may be preferable to those with low twist and high tilt.

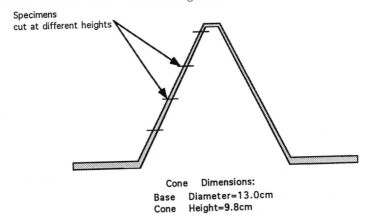

Cone   Dimensions:
Base   Diameter=13.0cm
Cone   Height=9.8cm

Figure 3: Section of Superplastic Cone

3. Specimens

Specimens were obtained from a cone test conducted on Ti-6Al-4V alloy at the superplastic temperature of $927°C$ and deformed at an equivalent strain rate of 2×10^{-4} per sec. Such a cone is designed to produce a state of balanced biaxial stress at all times [18]. Figure 3 shows a sectional view of the cone. Different strains were obtained at different heights of the cone. The strains were determined from the thickness of the specimen using $\epsilon = -\ln(t/t_0)$.

The specimens were mechanically polished using the 400-800 grit papers. Final polishing was done by using Chemomet solution. The final finish for the specimens was 0.05 microns. This fine finish was mandated by the BKD technique, and the relatively small grain size of the Ti-6Al-4V grains. During polishing, care was taken to avoid surface deformations. This was achieved by using small pressing forces (3-5 lbs) during polishing. It was not necessary to electropolish the specimens because there was no significant quality change in the BKD patterns for the electropolished test specimens.

Grain to grain orientation measurements were made for each of the strain levels.

Since it was not possible to get BKD pictures at each point, each grain was numbered so that it could be identified during the data analysis phase. The grains were enumerated by moving the specimen in straight directions; this was made possible by the micrometer screw gage attachments on the scanning electron microscope. The average grain size for the microstructure was about 2-4 microns. The grains were equiaxial for directions inside the normal plane. In table 1 the result of the study for the grain size distribution is presented.

4. Results

The direct pole figures for the basal planes are shown in figure 4. The pole figure for the initial configuration (0% strain) shows that all basal poles are aligned 30° from the normal direction. This is consistent with the reported results for low c/a ratio materials such as Titanium [9]. After 66% deformation all the poles are centered around the normal direction. One interesting feature of such results is that at 33% the (0001) basal fiber poles are spread away from the normal direction, however at 66% even though texture is reduced but the basal fiber texture become normalized.

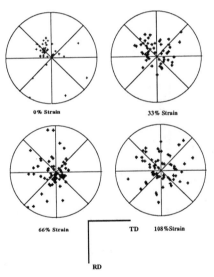

Figure 4: (0001) Pole Figures for 0%,33%,66%,108% specimens

A better understanding of the texture evolution is obtained by ODF analysis. A set of three Euler angles are used to coincide the crystallite coordinates and the principal directions of the material. The ODF's found using the BKD method are directly derived from the complete pole figure results of all random crystalline grains. The superplastically formed specimens are in a state of balanced biaxial stress. For the measurement of the pole figures and ODF the coordinate system was choosen such that only the normal to the specimen was maintained and the other two directions were choosen arbitrarily. Figure 6 shows the result of the derived ODF's for the initial structure and the consecutively strained specimens. In this case the distribution function is shown as a contour map on ψ,ϕ_1 axes with ϕ_2 held constant and is varied from 0 to 90 degrees with 10 degrees intervals. Every point in this plot represents a particular crystal orientation which can be expressed in terms of the ideal orientations (hklm) [uvtw]. In general, texture is reduced for all components. The basal texture near {0001} [2110] direction is about 13 times random for the unstrained specimen. The basal component of texture is reduced from 13 times random to 5 times random for 66% strain and stabilizes at 4 times random for strains of 108%. There is a weak component of (1011) texture near [2$\bar{3}$11] (two times random) in the initial material which is completely lost in the consecutive superplastic straining. Also the ($\bar{1}$013) component of texture shows a 4 times random near [2$\bar{1}\bar{1}$1] direction. This component reduces to 2 times random and stabilizes at that texture for the rest of the deformation process.

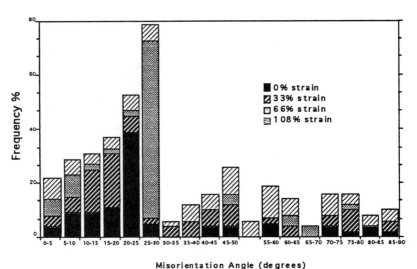

Figure 5: Frequency Distribution of Misorientation Angles

The result for $\langle 0110 \rangle$ component shows a high textured pole 35° from the normal direction. This cluster however completely disappears at the beginning of the super-plastic forming (33%)and aligns itself more in the TD and RD directions. A random distribution of this component is obtained after 66%. The results of the Misorientation distribution shows that for the original material exposed to the same temperature (annealing) conditions the is a high frequency of large and small misorientation angles

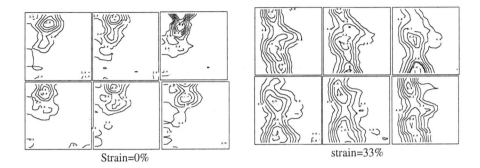

Strain=0%

strain=33%

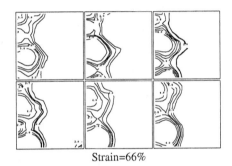

Strain=66%

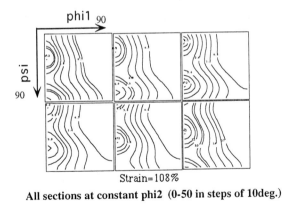

Strain=108%

All sections at constant phi2 (0-50 in steps of 10deg.)

Figure 6: ODF plots for 0% ,33%,66% and 108%

indicating the presence of grains and subgrains. Figure 5 shows the frequency plots of the misorientation angles of the four superplastic specimens.

Strain %	X-direction	Y-direction
0	2.3	2.6
33	3.5	3.0
66	3.4	3.3
108	3.5	3.2

Table 1: Grain Size Measurements (μm)

5. Conclusion

Texture studies invariably shows a reduction in the overall texture due to grain boundary sliding and grain rotation. In addition to the overall texture reduction, some texture components are stabilized and new ones are created indicating intergranular slip activity.

The purpose of the present study was to investigate the effect of superplastic deformation mechanisms on texture using the new technique of Backscattered Kikuchi Diffractometry. With the use of this technique not only the macroscopic texture can be measured, but also information on individual crystalline grain and grain rotations with respect to adjacent one become readily available. The addition of misorientation distribution information to the Orientation Distribution Function (ODF's) becomes very important in the analysis of micromechanisms of superplastic deformation since it gives a clear picture of the distribution of subgrain formation and their contribution in the final texture evolution.

References

[1] A.W. Bowen, *The Application of Texture Analysis to Studies of Superplastic Deformation* ,

[2] B.P. Kashayap, A.Arieli, A.K. Mukherjee, *Review of Microstructural aspects of superplasticity*, Journal of Materials Science, 20, 1985, pp 2661-2686.

[3] M. F. Ashby and R.A. Verrall,*Diffusion Accomodated Flow and Superplasticity* Acta Metta.,Feb. 91, pp 149-163

[4] T.G. Nien and J. Wadsworth, *Superplasticity and Superplastic Forming of Aluminium Metal-Matrix Composite*, Journal of Metals,Nov. 1992,pp 46-50

[5] I.L. Dillamore, W.T. Roberts Met. Rev. 10, No. 39,1965, pp 271.

[6] D. R. Thornburg and H.R. Piehler, *Cold-Rolling Texture Development in Titanium and Titanium Alloys.* Titanium. Proc. 2nd Int. Conf., Boston, MA, Plenum Press, New York, NY, 1973 pp.1187-93.

[7] C.J. Sparks, C.J. Hargue and J.P. Hammond, Transactions of Met. Soc. AIME 209, p. 49, 1957.

[8] V.V. Mukhayev, R. A. Adamesku and P.V. Geld (Sverdlovsk) Rus. Met. V, p. 67, 1968

[9] V. Venkatesan, S.T. Mahmood, and K.L. Murty, *Biaxial Creep Testing of Textured Ti-3Al-2.5V Tubing* Metallurgical Transactions A, Vol. 21A, November 1990, PP 3001-3010.

[10] *Superplasticity in Aerospace - Aluminium.* (Ed R Pearce and L. Kelly) Cranfield Inst. Tech (1985).

[11] K. Morii, C. Hartig, H. Mecking, Y. Nakayama, and G. Lutjering, Proceeding of 8th Int. Conf. on Texture of Materials

[12] K.N. Melton J. W. Edington, J. S. Kallend, *Textures in superplastic Zn-10% Al* Acta Metallurgica, Vol. 22, February 1974.

[13] R.J. Roe, J. Appl. Phys., 1965, vol. 36, p. 2024

[14] H.G. Bunge (Ed) *Experimental Techniques of Texture Analysis*, DGM Informationsgesellschaft Oberursel 1986.

[15] S. Matthies and W.G. Vinel : Phys. stat. sol. b112 (1982) K 111-120.

[16] S. Matthies : phys. stst. sol 92 (1972) k 135-138.

[17] J. Jura, K. Lucke and J. Pospiech : Z. Metallkde. 71 (1980) 714-728

[18] N. Chandra and D. Kannan,*Superplastic Sheet Metal Forming of a Generalized Cup*, Journal of Materials Engineering (in press), 1993.

[19] J. Zhao,*Analysis of misorientation in cubic crystals*,PhD Thesis, Brigham Young University, April,1988

AN INVESTIGATION OF HIGH STRAIN RATE SUPERPLASTIC

DEFORMATION MECHANISMS BY MEANS OF TEXTURE ANALYSIS

Zhe Jin and Thomas R. Bieler

Department of Materials Science and Mechanics
Michigan State University, East Lansing, MI 48824

Abstract

Texture analysis is useful for investigation of superplastic deformation behavior since effects from the individual deformation mechanisms, such as grain boundary sliding (GBS), dislocation slip, recrystallization, etc. can be separated. We have investigated high rate superplastic deformation in mechanically alloyed aluminum IN90211. Three specimens deformed at 475°C in regions I, II and III were analysed using Sample Orientation Distribution Functions (SODF). The texture evolution with strain was analyzed and deformation mechanisms in each regime are discussed in terms of how the texture changed. Initial (undeformed) textures were composed of two skeletons in all three specimens, and their configurations were generally maintained throughout deformation, but the changes in intensities and locations differed in the three specimens. The results indicate that in region I, GBS is the dominant deformation mechanism, but dislocation slip also contributes to the deformation; in region II, the superplastic deformation occurs by GBS, dislocation slip, and recrystallization (at larger strains); in region III, the dominant contribution to deformation is from dislocation slip. These results indicate that high rate superplastic deformation results from an optimal balance between dislocation slip and GBS.

Advances in Superplasticity and Superplastic Forming
Edited by N. Chandra, H. Garmestani, R.E. Goforth
The Minerals, Metals & Materials Society, 1993

Introduction

Superplastic deformation has been extensively studied by various physical metallurgy methods using optical and electron microscopy (1-4). The deformation mechanisms commonly explored in superplasticity have been grain boundary sliding, dislocation slip, dynamic recrystallization, and diffusional creep. Texture analysis has been effectively used for investigating crystal deformation behaviors (5,6), including superplasticity (7). However, the use of texture in the study of superplastic deformation, particularly at high strain rates, (8,9) is limited. Individual texture components can be followed throughout the deformation history to clarify the role of different deformation mechanisms. Prior results have shown that grain boundary sliding smears (weakens) existing components of texture, but does not eliminate the identity of these components. Dislocation slip and dynamic recrystallization create specific texture components during deformation (6,8,9).

High strain rate (positive exponent) superplasticity, hereafter denoted PES, has been found in a number of materials (10-12). In general, PES is found in materials containing small particles or whiskers, which keep the grain size small. Superplastic deformation mechanisms of IN90211 mechanically alloyed aluminum have been analyzed in detail (10). At the stress and strain rates where most superplastic materials exhibit optimum elongations, PES materials exhibit low elongations. At a sufficiently large threshold stress, the strain rate sensitivity increases rapidly to values typical of superplasticity. Due to the larger stresses needed for PES the role of slip may differ compared to conventional superplaticity. To explore this, we have studied the deformed specimens of IN90211 from (10). Since the deformed specimens exhibit a diffuse neck, there is a range of strains represented in one specimen. We have measured the textures at several positions to determine how texture varies with strain.

Experimental Procedures and Results

The composition of IN90211 mechanically alloyed aluminum is given in Table I. The material was extruded from mechanically alloyed powders, forged and rolled at elevated temperature, and annealed at 492°C for 1 hour and water quenched prior to machining 6.5 mm long tensile specimens parallel to the rolling direction. Three tensile specimens deformed at 475°C were investigated; $\dot{\epsilon} = 1 \times 10^{-4}$ sec^{-1} (44% elongation) in region I; $\dot{\epsilon} = 77$ sec^{-1} (400% elongation) in region II, and $\dot{\epsilon} = 330$ sec^{-1} (70% elongation) in region III. Further experimental detail can be found in (4,10). At the deformation temperature it is probable that most of the Al$_2$Cu had dissolved.

Table I   Composition (wt%) of IN90211

	Mg	Cu	C	O	Al$_2$Cu	Al$_4$C$_3$	Al$_2$O$_3$	Al
	2.0	4.4	1.1	0.8				91.7
*	2.0	0.9			6.4	4.3	1.7	84.7
*					4.1vol%	4.1vol%	1.2vol%	90.6vol%

* Assuming 80% of Cu in Al$_2$Cu, O and C in Al$_2$O$_3$ and Al$_4$C$_3$, and Mg in solution.

Texture measurements were made using X-ray diffraction with Cu radiation. Pole figures were measured at several positions along the center of the specimen with an exposure area of 1.5mm diameter at each point and in the undeformed shoulder region. The sample normal was parallel to the rolling plane normal. The true strain ε at each point was determined from the reduction of area by assuming constant volume during deformation. Near the fracture surface, the final strain rate was higher than the nominal rate. The (111), (200) and (220) pole figures measured at each point were used to create orientation distribution functions (ODFs) using the preferred orientation package from Los Alamos (popLA) using procedures described in (13). The data were corrected with an analytical defocusing curve rather than experimental curves, so the data were analysed in a comparative manner, normalized to the undeformed part of the specimen. The sample orientation distribution functions (SODFs) were calculated using a series expansion of spherical harmonics with expension $L_{max}=22$.

All specimens exhibited textures with two skeletons that result from the prior deformation processing. Since the initial texture varies slightly from specimen to specimen, analysis was done by measuring changes from the initial undeformed texture. The changes in intensity and location of these skeletons were followed as a function of strain.

Results and Analysis

Since ODFs are advantageous compared to pole figures for their unique description of texture, we present the texture using results from the sample orientation distribution function (SODF) using Bunge's notation. The results for region II will be described after regions I and III since region II is the most complex.

The texture evolution during superplastic deformation is related to grain boundary sliding (GBS), grain rotation, dislocation slip, and recrystallization (8,9,14,15). Dislocation slip in fcc crystals can be divided into single slip (one slip system operating), double slip (two slip systems operating) and multiple slip (more than two slip systems operating). The single slip and double slip usually cause ⟨112⟩ fiber texture, while multiple slip results in ⟨110⟩, ⟨111⟩ and ⟨100⟩ fibers depending on the number of operating slip systems. ⟨110⟩ texture results from the operation of four slip systems, ⟨111⟩ is from six systems and ⟨100⟩ is from eight. Grain boundary sliding (GBS) is considered the significant superplastic deformation mechanism. Since GBS depends upon boundary misorientation angle but not grain orientation, its operation generally reduces existing texture intensity uniformly (7). Recrystallization mechanisms predict a strong cube texture since cube texture has favored orientation relationship to deformed matrix for the nucleation and rapid growth of recrystallized grains (16). Using these results, the operative mechanisms in the current work can be determined.

(a) Textures and Deformation in Region I ($\dot{\epsilon}=1\times10^{-4}$/sec)

Texture is mainly composed of two skeleton lines that remain intact throughout the deformation. The arrangement of these two skeletons in Euler space is shown in Fig. 1. Strain alters the intensity and the position of the skeleton. Fig. 2 shows the intensity change of two skeletons as a function of strain. It indicates that the orientation densities along the skeletons decrease with strain and reach the lowest points at true strain $\varepsilon=1.13$, near the fracture surface. The magnitude of deviation of skeletons from the initial skeleton line are plotted in Fig. 3, and denoted as the shift angle (direction of shift is known, but omitted to simplify the description). Skeleton 1 shifts with increasing strain, but only a minor shift is observed in skeleton 2 near the fracture point.

In the SODF of Fig. 1, the overall intensity of texture at each strain decreases dramatically with increasing strain. This overall intensity drop indicates that a significant amount of GBS occurred during deformation. But since cavitation occurs (4), the high elongations associated

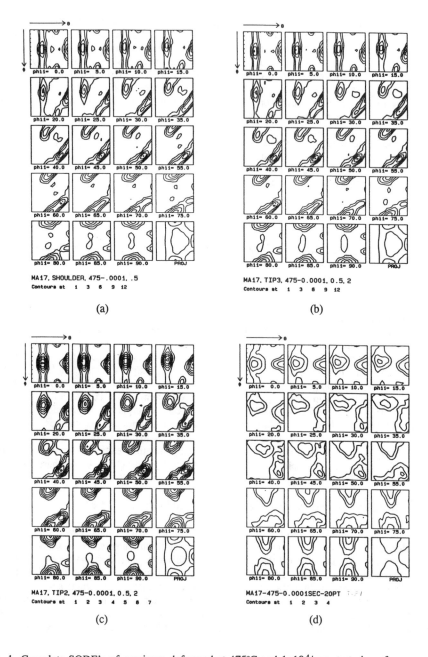

Fig. 1 Complete SODF's of specimen deformed at 475°C and 1x10⁻⁴/sec at strains of
(a) 0; (b) 0.45; (c) 0.72; (d) 1.38.

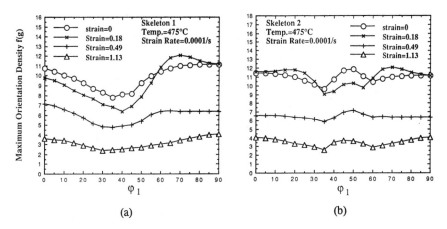

(a)   (b)

Fig. 2 Intensity change of skeletons with strain (a) in skeleton 1 and (b) in skeleton 2.

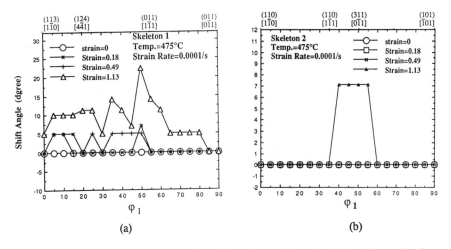

(a)   (b)

Fig. 3 Skeleton line shifts from the initial skeleton line with strain (a) in skeleton 1 and
(b) in skeleton 2.

with this deformation mechanism are precluded. In this region, the GBS superplastic deformation conditions are satisfied, but the low strain rate sensitivity and the rapid cavitation rate precludes superplastic elongation. Bieler's analysis (10) indicates that the grain boundary sliding process depends upon the rate of arrival of lattice dislocations in the grain boundary (as extrinsic boundary dislocations). Since dislocation density is proportional to σ^2 the stress may be insufficient to permit enough lattice dislocations to accommodate stress concentrations in the boundaries. The threshold stress behavior observed at the transition between regions I and II describes the stress where enough accommodations occurs on all boundaries to prevent cavitation.

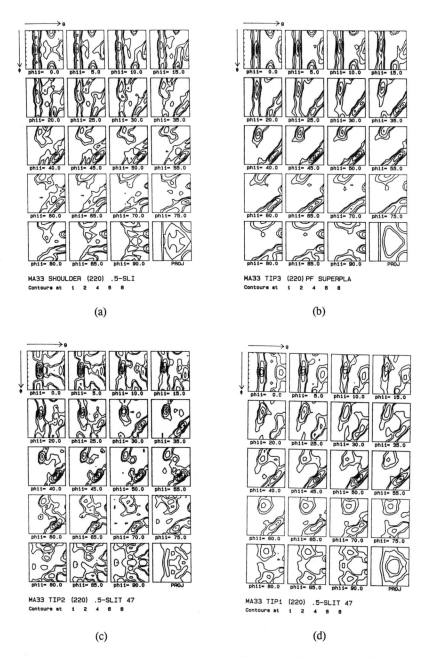

Fig. 4 Complete SODF's of specimen deformed at 475°C and 330/sec at strain of (a) o;
(b) 0.45; (c) 0.72; (d) 1.38.

In Fig. 3, the shift angles of skeletons increase with increasing strain. This indicates that dislocation slip occurred systematically during deformation, since dislocation slip will rotate the existing texture components away from their initial locations (17). Skeleton 1 shifts much more than the skeleton 2, indicating that the grains contributing to skeleton 2 are not rotating so much. In skeleton 1, the shift angle rises dramatically close to the fracture surface ($\varepsilon = 1.13$). This indicates that more dislocation slip occurred in the neck region. This region was straining at a higher stress and strain rate than given by the nominal strain rate.

(b) Textures and Deformation in Region III ($\dot{\varepsilon} = 330/sec$)

Fig. 4 shows the complete SODF's of texture in region III. As in region I, it contains two skeletons but the path of the skeleton line differs slightly from the region I specimen. The intensity changes with strain of the two skeletons are shown in Fig. 5. As strain increases in

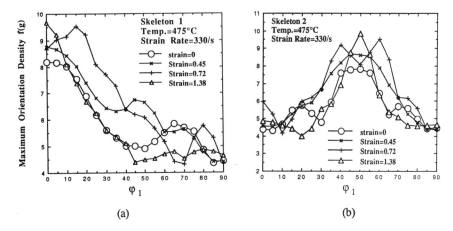

(a)   (b)

Fig. 5   Intensity changes of skeletons with strain (a) in skeleton 1 and (b) in skeleton 2.

Fig. 5(a), a peak of the skeleton line moves towards low φ_1 angle. The general shapes of texture contours in SODF's do not change much with strain, but the skeleton line shifts from its original location at $\varepsilon = 0$ as shown in Fig. 6. The skeleton line does not shift uniformly from the original line but the monotonic shift with increasing strain is obvious in Fig. 6 for both skeleton lines.

Texture in this region is much different than that in region I. The maximum texture intensity steadily increases with strain, indicating that the dominant deformation mechanism is not GBS. The shifts are larger in magnitude than those in region I (Fig. 3), indicating that dislocation slip is the significant and dominant mechanism. The magnitude of the shift with strain along the two skeletons are almost the same (Fig. 6a,b). This indicates that grains in both skeletons rotate similarly during deformation in region III. Unlike the region I situation, rate of shift with strain decreases for both skeletons at the neck. This suggests that another deformation mechanism, such as GBS, operates in addition to the dislocation slip. If GBS occurs during the necking, it may result from adiabatic heating, which would accelerate the kinetics for sliding (10).

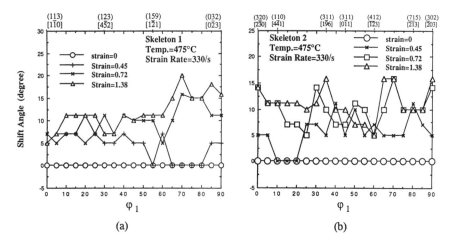

Fig. 6 Skeleton line shifts from the initial skeleton line with strtain (a) in skeleton 1 and (b) in skeleton 2.

(c) Textures and Deformation in Superplastic Region II ($\dot{\varepsilon} = 77$/sec)

In this case, the texture evolution with strain is much more complicated. In addition to the two skeletons observed above, there is also a strong ⟨012⟩//ND fiber texture. Fig. 7 shows the SODF's of texture in region II. Taking [021]//ND for example, it runs from (021)[100] to (021)[012] with the [021] crystal direction constantly perpendicular to the sample rolling plane. The intensity changes of the two skeletons and the [021]//ND fiber are presented in Fig. 8. In Fig. 8(c), the intensities are measured along an exact fiber orientation of ($\varphi_1, 63°, 0°$) for all strains.

The shift angles of the two skeletons with strain do not show a monotonically changing trend like the previous two specimens (in Fig. 9). The maximum shift occurs at a strain of $\varepsilon = 2.13$. At the next larger strain, the shift decreases. In addition, cube texture {100}⟨010⟩ appears at $\varepsilon = 2.13$, with an intensity of 1.71.

The overall intensity decreases with strain at lower strains reaching a minimum value and increases significantly at the strain close to the fracture point (see Fig. 7). The decrease indicates that GBS initially dominates the deformation, but dislocation slip also occurs; in Fig. 9 the shift angles increase with strain during deformation. Therefore, dislocation slip contributes to the total deformation, but it is less dominant than GBS.

As the neck forms the increasing stress increases the dislocation slip activity. This results in the increase of overall intensity near the fracture (Fig. 7(d)). One significant difference from the previous cases in Fig. 9 is that the shift angles become very large at $\varepsilon = 2.13$ for both skeletons and then the shift decreases again. The large, sudden shift results from the appearance of a cube component (Fig. 7c), which could arise from recrystallization. However, the cube orientation is not stable with further strain, and rotations in all directions are possible, so continuing deformation can quickly rotate the cube components away. The result is that the shift angles for both skeletons decrease from their peak values in Fig. 9 with continuing

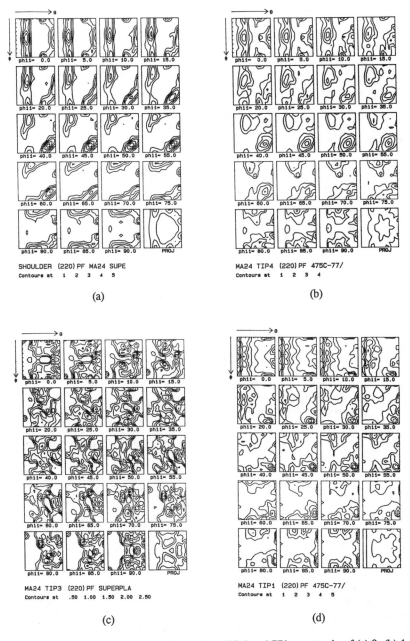

Fig. 7 Complete SODF's of specimen deformed at 475°C and 77/sec at strain of (a) 0; (b) 1.6; (c) 2.13; (d) 2.86.

strain but they still have the values larger than those at lower strains.

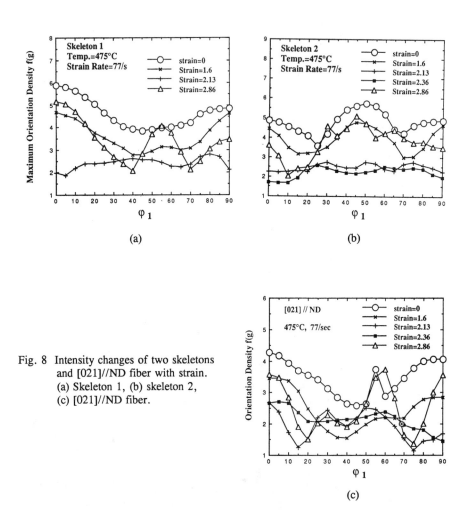

Fig. 8 Intensity changes of two skeletons
and [021]//ND fiber with strain.
(a) Skeleton 1, (b) skeleton 2,
(c) [021]//ND fiber.

Discussion

The description of orientation changes was made using simple principles associated with dislocation slip, recrystallization, and grain boundary sliding in monolithic materials. Even though this material has about 6 vol% particles that may alter deformation behavior, a self consistent view of deformation in this material is obtained. The texture analysis provides information that can compliment interpretations made from microscopy. Whereas TEM investigations represent a statistically small volume of the material, each texture measurement performed in this study sampled over a million grains. A description of deformation consistent with both TEM and texture evolution will bring confidence to any model.

This study indicates that slip during PES deformation is significant in this alloy. Though not analysed in this paper, it is possible to esitmate the relative amounts of strain resulting from slip and GBS from these data. Qualitatively, deformation in region I exhibits slip, but the

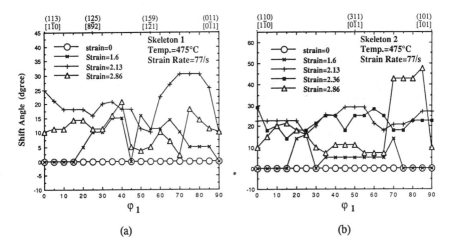

Fig. 9 Skeleton shifts from the initial skeleton line with strain (a) in skeleton 1 and (b) in skeleton 2.

randomization from GBS is stronger than in region II and III samples. GBS is unimportant in region III. The observation that more slip occurred in region II than region I differs from the common understanding of superplasticity, where grain boundary sliding is usually maximum in region II. In this material, GBS may be maximum in region II, but the concurrent slip counteracts the randomizing effects of GBS. Therefore slip in PES deformation is more important than in conventional superplastic materials.

PES materials tend to have particles or fibers in the microstructure that keep the grain size small enough for PES to occur (1). However, the particles also generate threshold stresses for dislocation slip (10). Since motion of grain boundary dislocations is geometrically necessary for GBS, a threshold stress for motion of grain boundary dislocations can account for the resistance to high elongation at conventional strain rates. Since boundary dislocation motion is inhibited in region I, and since vacancy diffusion rates are high, cavity nucleation and growth can occur and accelerate fracture during the experiment. Particles also increase the stress exponent and thus flow instabilities are also accelerated. Even though grain boundary sliding is dominant, the cavity growth and flow instability prevent superplatic elongations. At higher stresses and strain rates, where grain boundary dislocations can overcome obstacles and the time is reduced for cavity growth, the GBS process occurs without cavitation and with a lower stress exponent. In addition, the higher stress increases the mobile dislocation density in the grains (10).

Conclusions

(1) In region I and region III, most orientations are located in two skeletons. In region II, cube texture appears at strain $\varepsilon=2.13$ and the [021]//ND fiber are also present.

(2) Deformation in region I is dominated by GBS, but dislocation slip also contributes to the deformation. The low elongation results from caviatation and the low strain rate sensitivity caused by threshold stresses resulting from the particles in the alloy. Once the stress exceeds the threshold stress, resistance to GBS stops, and superplastic elongations are obtained.

(3) Deformation in region II is due to a combination of GBS and dislocation slip, but the contribution of GBS affects the texture intensity changes more than dislocation slip. Recrystallization occurred at $\varepsilon=2.13$ and made a large contribution to texture changes at that strain.

(4) Deformation in region III is completely dominated by dislocation slip.

Acknowledgements

The specimens used in this study were taken from research funded by AFOSR-86-0091.

References

(1)   O.D. Sherby and J. Wadsworth, Prog. Mater. Sci., 33 (1989), 169.
(2)   B.P. Kashyap, A. Arieli and A.K. Mukherjee, J. Mater. Sci., 20 (1985), 2661.
(3)   K. Matsuki, et.al., Z. Metallkde, 79, (1988), 231.
(4)   T.R. Bieler, G.R. Goto and A.K. Mukherjee, J. Mat. Sci., 25 (1990), 4125.
(5)   J. Hirsch and K. Lucke, Acta Metall., 33 (1985), 1927.
(6)   K. Lucke, ICOTOM7, International Conference on Texture of Materials, eds. C.M. Brakman, P. Jongenberger, E.G. Mittemeijer, Netherlands Society for Materials Science, (1984), 195-210.
(7)   K.A. Padmanabhan and K. Lucke, Z. Metallkde 77, (1986), 765.
(8)   Z. Jin and T.R. Bieler, Proceedings on Superplasticity in Advanced Materials, ed. by S. Hori, M. Tokizane and N. Furushiro, The Japan Society for Research on Superplasticity, 1991, p. 587.
(9)   Z. Jin and T.R. Bieler, Texture Change During Superplastic Deformation of Mechanically Alloyed Aluminum IN90211, J. Mat. Sci., in press.
(10)   T.R. Bieler and A.K. Mukherjee, Mater. Sci and Eng. A128 (1990) 171-182.
(11)   M. Mabuchi et. al., Scripta Metall. et. Mater. 25, (1991) 2517
(12)   K. Higashi et. al., Scripta Metall. et. Mater. 26, (1992) 185.
(13)   J.S. Kallend, U.F. Kocks, A.D. Rollet, H.-R. Wenk, Mater. Sci. and Eng. A132, (1991), 1-11.
(14)   Z. Jin and T.R. Bieler, unpublished results.
(15)   C.P. Cutler, J.W. Edington, J.S. Kallend and K. N. Melton, Acta Metall. 22 (1974), 665.
(16)   I.L. Dillamore and H. Katoh, Metal Sci., 8 (1974), 73.
(17)   R.W.K. Honeycombe, The Plastic Deformation of Metals, Edward Arnold, 1984.

NUMERICAL MODELING OF 3-D SUPERPLASTIC

SHEET FORMING PROCESSES

S. C. Rama*, N. Chandra *and* R. E. Goforth*

Department of Mechanical Engineering, FAMU/FSU College of Engineering
Florida State University, Tallahassee, FL 32316-2175

*Department of Mechanical Engineering
Texas A&M University, College Station, TX 77843

Abstract

SuperPlastic Forming (SPF) is widely accepted as an advanced technique for forming complex industrial components. To achieve optimum superplasticity (maximum ductility) the deformation process has to be carried out with a precise control on maximum true strain-rate. The controlling strain-rate control can be constant or varying depending on whether the material is statically recrystallized or dynamically recrystallizing respectively. Hence for economical and successful application of SPF to industrial components there is a great need for modeling the forming process. A three dimensional thin shell formulation is presented in this paper. The formulation uses a constant stress constant moment triangular thin shell element using a convective coordinate system. The contact algorithm is incorporated through a pseudo equilibrium concept whereas the time integration is performed using an explicit scheme. Results of this 3-D model are compared to experimental data and results from two dimensional finite element and simpler membrane models.

Advances in Superplasticity and Superplastic Forming
Edited by N. Chandra, H. Garmestani, R.E. Goforth
The Minerals, Metals & Materials Society, 1993

INTRODUCTION

Superplasticity in materials is achieved only in a narrow range of strain-rate with an optimum value unique to each material[1]. Hence in the design of complex structural components using SPF processes, it is necessary to determine the pressure loading cycle to maintain the true strain-rate within the narrow range, during the entire forming process. Hence for the economical and successful application of SPF to industrial components there is a great need for modeling. The finite element method (FEM) is used both as an analysis and a design tool [2-11]. Extensive work has been done in modeling two dimensional and axisymmetric configurations using continuum formulations [2-5]. To handle complex 3-D configurations, the continuum formulation is found to be computationally intensive. Many three dimensional finite element models have been developed by treating the superplastic sheet as a Newtonian viscous membrane [6-11]. A three dimensional thin shell approach to modeling superplastic forming is presented in this paper. The finite element process models are used to determine the process parameters which include the pressure-time loading and final thickness distribution. The determination of these process parameters is very critical in not only reducing the forming time but also in forming complex components without fracture.

Superplastic materials exhibit the ability to have very large neck-free elongations (superplastic property) only under certain forming conditions of temperature and strain-rate. Once formed into complex shapes at high temperatures the superplastic component is structurally strong at the operating (near room) temperature with a yield strength many times greater than that when formed. The uniaxial flow stress σ of a superplastic material is a strong function of the inelastic strain-rate $\dot{\epsilon}$ and a weak function of the strain ϵ and grain size d. From a mechanistic point of view, a constitutive model can be postulated as [1],

$$\sigma = K_1 \dot{\epsilon}^m \epsilon^n d^p \tag{1}$$

where K_1, m, n, and p are material constants. Based on experimental observations, the most common form of the constitutive equation used is the classical power-law material model given by

$$\sigma = K \dot{\epsilon}^m \tag{2}$$

where K is a material constant and m is the strain-rate sensitivity. Many other forms of uniaxial equations have also been suggested, and excellent reviews of the effect of different material parameters on flow stress are given in Reference[1].

FORMULATION

For the thin shell finite element analysis adapted in this paper, the nonlinearities arise due to complex die geometry, material, nonconservative pressure loading, and contact. Also when FEM is used as a design tool the forming pressure (kinetic quantity as the input) required to maintain optimum strain-rate (a kinematic quantity as output) needs to be determined which is the reverse of input/output quantities in a conventional method.

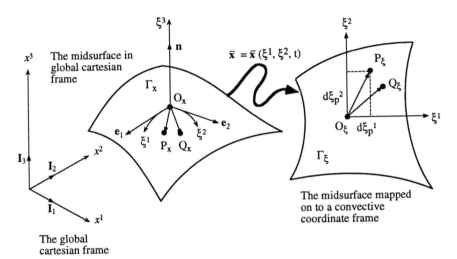

Figure 1. Definition of coordinate frames and midsurface geometry

Geometry and Kinematics

Since in superplastic forming, transverse shear and bending do not play a significant role when compared to the membrane effects. Kirchhoff thin shell formulation which neglects transverse shear effects is adapted in this work. The Kirchhoff theory effectively models the required membrane and bending response at a less computational price. The shell mechanics explained in this section is based on a set of convective coordinate system (ξ^α) that conforms to the midsurface of the deforming shell. The in-plane reference coordinate directions (ξ^1, ξ^2) remain perfectly tangent to a given set of material lines embedded on the midsurface of the deforming shell; and the out-of-plane coordinate axis ξ^3 remains normal to the plane of the midsurface at all locations and at all times (Figure 1).

The midsurface of the shell can be described by a mapping from each particle (ξ^1, ξ^2, ξ^3) to global Cartesian $\Re^3$ as:

$$\mathbf{x}(\xi^1, \xi^2, \xi^3, t) = \bar{\mathbf{x}}(\xi^1, \xi^2, t) + \frac{h}{2}\xi^3 \mathbf{n}(\xi^1, \xi^2, t) \tag{3}$$

where the symbol *over bar* ($^-$), denotes the value of any quantity ( ) associated with the midsurface. h represents the thickness of the shell and $\mathbf{n}$ the unit normal to the midsurface.

The covariant base vectors described with respect to convective system can be described as,

$$\mathbf{e}_\alpha = \frac{\partial \mathbf{x}}{\partial \xi^\alpha} = \bar{\mathbf{e}}_\alpha + \frac{h}{2}\frac{\partial \mathbf{n}}{\partial \xi^\alpha} \qquad \alpha = 1, 2$$

$$\mathbf{e}_3 = \frac{h}{2}\mathbf{n} \tag{4}$$

The convective bases $\bar{\mathbf{e}}_\alpha, \alpha = 1, 2$ is associated with the midsurface and is always tangent to the midsurface, and $\mathbf{e}_3 = \bar{\mathbf{e}}_3$ is always normal to the midsurface.

135

The metric tensor can be derived as,

$$g_{\alpha\beta} = \mathbf{e}_\alpha \cdot \mathbf{e}_\beta = \bar{g}_{\alpha\beta} + \frac{h}{2}\xi^3 \bar{\kappa}_{\alpha\beta} \qquad \alpha, \beta = 1, 2, 3 \tag{5}$$

where $\bar{g}_{\alpha\beta} = \bar{\mathbf{e}}_\alpha \cdot \bar{\mathbf{e}}_\beta$ is the metric tensor associated with the midsurface whereas $\bar{\kappa}_{\alpha\beta}$ is described as the curvature of the midsurface. Defining curvature as the variation of normal vector throughout the sheet, the covariant convective components of the curvature tensor are given as,

$$\bar{\kappa}_{\alpha\beta} = -\frac{\partial \mathbf{n}}{\partial \xi^\alpha} \cdot \bar{\mathbf{e}}_\beta = \mathbf{n} \cdot \frac{\partial^2 \bar{\mathbf{x}}}{\partial \xi^\alpha \partial \xi^\beta} \qquad \alpha, \beta = 1, 2 \tag{6}$$

The contravariant components of the metric tensor can be derived using the contravariant basis similar to the derivation of the covariant components in equation (5)

$$g^{\alpha\beta} = \mathbf{e}^\alpha \cdot \mathbf{e}^\beta = \bar{g}^{\alpha\beta} + \frac{h}{2}\xi^3 \bar{\kappa}^{\alpha\beta} \qquad \alpha, \beta = 1, 2, 3 \tag{7}$$

Since $\mathbf{e}_\alpha, \alpha = 1, 2$ are normal to $\mathbf{e}_3$ we can derive that,

$$g_{\alpha 3} = \mathbf{e}_\alpha \cdot \mathbf{e}_3 = 0 \qquad \alpha = 1, 2; \qquad g_{33} = \mathbf{e}_3 \cdot \mathbf{e}_3 = \frac{h^2}{4} \tag{8}$$

It will be seen later that the metric tensor $\bar{g}_{\alpha\beta}$ gives a measure of the in-plane deformation of the midsurface and the quantity $h\xi^3 \bar{\kappa}_{\alpha\beta}$ represents the out-of- plane deformation of the shell relative to the midsurface. This deformation is accommodated by the curvature of midsurface of the shell.

Taking the time derivative of the equation of motion, given in equation (3), enables the velocity of any given particle having reference coordinates $\{\xi^\alpha\}$ to be expressed as,

$$\mathbf{v}(\xi^1, \xi^2, \xi^3, t) = \frac{\partial \mathbf{x}}{\partial t} = \bar{\mathbf{v}}(\xi^1, \xi^2, t) + \frac{h}{2}\xi^3 \dot{\mathbf{n}}(\xi^1, \xi^2, t) \tag{9}$$

The covariant convective components of the rate of deformation tensor $\mathbf{D}$ can be derived as

$$D_{\alpha\beta} = \frac{1}{2}\dot{g}_{\alpha\beta} = D^M_{\alpha\beta} + \frac{h}{2}\xi^3 D^B_{\alpha\beta} \qquad \alpha, \beta = 1, 2 \tag{10}$$

$$D_{\alpha 3} = 0 \qquad \alpha = 1, 2; \qquad D_{33} = \frac{1}{4}h\dot{h} \tag{11}$$

where the convective components of the membrane rate of deformation tensor $\mathbf{D}^M$ is given by,

$$2D^M_{\alpha\beta} = \dot{\bar{g}}_{\alpha\beta} = \frac{\partial \bar{x}^i}{\partial \xi^\alpha} \frac{\partial \bar{v}^i}{\partial \xi^\beta} + \frac{\partial \bar{x}^i}{\partial \xi^\beta} \frac{\partial \bar{v}^i}{\partial \xi^\alpha} \qquad \alpha, \beta = 1, 2 \text{ and } i = 1, 2, 3 \tag{12}$$

where the components with α and β are referred to the convective system while the index i denotes the components referred to the global Cartesian system. The convective covariant component of the bending strain-rate tensor, $\mathbf{D}^B$ can be simply derived as,

$$D^B_{\alpha\beta} = \dot{\bar{\kappa}}_{\alpha\beta} = \mathbf{n} \cdot \frac{\partial^2 \bar{\mathbf{v}}}{\partial \xi^\alpha \partial \xi^\beta} = \frac{\partial^2 \dot{w}}{\partial \xi^\alpha \partial \xi^\beta} \tag{13}$$

where $\dot{w} = \bar{\mathbf{v}} \cdot \mathbf{n}$ is the normal velocity. Detailed derivations for these expressions are presented in Reference [12].

Constitutive Equations

The power law model cast cast by expressing equivalent stresses as a function of the equivalent strain-rate is found to be an effective constitutive relation in modeling superplastic behavior. The equation is given by,

$$\bar{\sigma} = K\dot{\bar{\epsilon}}^m \tag{14}$$

Thus the superplastic alloys can be modeled as a non-Newtonian incompressible viscous fluid, wherein the deviatoric stress tensor $\hat{\sigma}'$ is related to the strain-rate tensor as

$$\hat{\sigma}' = 2\mu(\mathbf{D})\mathbf{D} \tag{15}$$

where μ is the nonlinear viscosity and $\mathbf{D}$ is the strain-rate tensor. Using the definitions of $\bar{\sigma} = \sqrt{\frac{3}{2}\hat{\sigma}' : \hat{\sigma}'}$ and $\dot{\bar{\epsilon}} = \sqrt{\frac{2}{3}\mathbf{D} : \mathbf{D}}$ in equation (14), and using the relationship in equation (15), the equivalent stress can be written as,

$$\bar{\sigma} = 3\mu\dot{\bar{\epsilon}} \tag{16}$$

Comparing equations (14) and (16) the nonlinear viscosity can be defined as

$$\mu = \frac{K}{3}\dot{\bar{\epsilon}}^{m-1} \tag{17}$$

It is assumed that the sheet behaves like a membrane with a plane stress condition across the thickness. Thus $\sigma^{33} = 0$, and incorporating the incompressibility constraint expressed as, $g^{\gamma\delta}D_{\gamma\delta}$, $\gamma, \delta = 1, 2, 3$, the constitutive equations, in terms of the convective membrane stress resultants and the rate of deformation can be written as

$$\sigma^{\alpha\beta} = 2\mu G^{\alpha\beta\gamma\delta} D_{\gamma\delta} \tag{18}$$

where $G^{\alpha\beta\gamma\delta}$ is related to the metric tensor as

$$G^{\alpha\beta\gamma\delta} = g^{\alpha\beta}g^{\gamma\delta} + g^{\alpha\gamma}g^{\beta\delta} \tag{19}$$

Substituting equation (10) in equation (18) the membrane stress resultants and the bending resultants can be derived as follows [12],

$$\sigma_M^{\alpha\beta} = 2\mu^M \, h \, G^{\alpha\beta\gamma\delta} \, D_{\gamma\delta}^M \tag{20}$$

$$\sigma_B^{\alpha\beta} = 2\mu^B \, \frac{h^3}{12} \, G^{\alpha\beta\gamma\delta} \, D_{\gamma\delta}^B \tag{21}$$

<u>Equilibrium Equations</u>

Multiplying the equilibrium equation with virtual velocities and integrating over domain of volume V,

$$\int_V (\nabla \cdot \hat{\sigma} + \mathbf{f}) \delta \mathbf{v} \; dV = 0 \tag{22}$$

After integrating by parts and applying the Gauss divergence theorem we have,

$$\int_V \hat{\sigma} : \delta \mathbf{D} \; dV = \int_V \mathbf{f} \cdot \delta \mathbf{v} \; dV + \int_\Gamma \mathbf{t} \cdot \delta \mathbf{v} \; dA \tag{23}$$

where the left hand side of the equation is the internal rate of virtual work and the right hand side corresponds to the external rate of virtual work. Now the rate of deformation tensor can be decomposed into its membrane and bending parts and hence the internal rate of virtual work can be written as,

$$\begin{aligned} \delta \dot{W}_{\text{int}} &= \delta \dot{W}_{\text{int}}^M + \delta \dot{W}_{\text{int}}^B \\ &= \int_{m\Gamma} \sigma_M^{\alpha\beta} \; \delta D_{\alpha\beta}^M \; d\bar{A} + \int_{m\Gamma} \sigma_B^{\alpha\beta} \; \delta D_{\alpha\beta}^B \; d\bar{A} \qquad \alpha, \beta = 1, 2 \end{aligned} \tag{24}$$

The external rate of virtual work consists of the body forces and surface tractions. The surface tractions comprise of normal or tangential forces and also moments applied to the midsurface. Neglecting the body forces and the moments for the surface traction, the external rate of virtual work can be expressed in terms of the normal and tangential forces as,

$$\delta \dot{W}_{\text{ext}} = \int_{m\Gamma} p^i \delta \bar{v}^i \; d\bar{A} \tag{25}$$

<u>Finite Element Discretization</u>

The simple Kirchhoff shell element considered in this formulation is a constant moment, constant stress element and has 3 corner nodes and 3 midside nodes. Each of the three corner nodes have translational velocities as degrees of freedom and hence are greatly responsible for the geometric description of the element. The 3 midside nodes have 1 rotational degree of freedom each. Hence the geometry of the shell can be described by the three corner nodes through shape functions $N_A(\xi^1, \xi^2)$ (see Figure 2) as,

$$\bar{\mathbf{x}}(\xi^\alpha, t) = \sum_{A=1}^{n_{np}} \bar{\mathbf{x}}_A(t) \; N_A(\xi^\alpha) \qquad \alpha = 1, 2 \tag{26}$$

where n_{np} represents the number of nodes ($=3$) in the element considered. The *overbar* signifies that the quantities considered are associated only with the shell midsurfaces. The velocities associated with the midsurface $\bar{\mathbf{v}}$ can be represented as,

$$\bar{\mathbf{v}} = {}^t\bar{\mathbf{v}} + \dot{w}\mathbf{n} \tag{27}$$

where ${}^t\bar{\mathbf{v}}$ are the tangent and $\dot{w}\mathbf{n} = (\mathbf{n} \cdot \bar{\mathbf{v}})\mathbf{n}$ are the normal velocity vectors to the midsurface of the shell. The shape functions N_A are also used to interpolate the in-plane (or tangent) velocities of the shell element as,

138

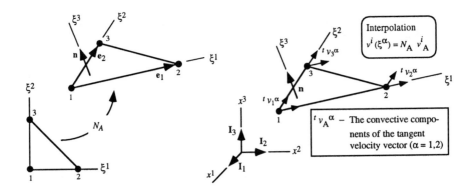

Figure 2. Description of the triangular element in the convective coordinate
system with interpolation functions

$$^t\bar{\mathbf{v}}(\xi^\alpha) = \sum_{A=1}^{n_{np}} {}^t\bar{\mathbf{v}}_A \, N_A(\xi^\alpha) \qquad \alpha = 1,2 \tag{28}$$

whereas the normal velocity is interpolated as,

$$w(\xi^1, xi^2) = w_A\psi_A(\xi^1,\xi^2) + \dot{\theta}_{A'}\phi_{A'}(\xi^1,\xi^2) \qquad A = 1,2,3 \text{ and } A' = 1,2,3 \tag{29}$$

where the ψ_A and ϕ_A are the shape functions for the transverse velocities; A and
A' denotes the corner and midside nodes respectively. Expression for these shapes
functions are derived in Reference [12].

Even though the midside nodes contribute in providing a curvature to this otherwise flat element, it should be noted that this element is not C^1 continuous element.
As a result of the limited number of degrees of freedom per element, neither the velocity field across the edges (along the midside nodes) nor the rotational velocities at
the corner nodes are continuous. Thus the relationships derived above hold good only
inside of each element.

Substituting these shape function in the kinematic equations and then substituting
the constitutive relationship in the equilibrium equations the internal equivalent nodal
forces comprise of the membrane $^M\mathbf{F}_A$ and bending $^B\mathbf{F}$ parts as well as moments.
Considering these internal forces over one element e, we have,

$$^{M,e}\mathbf{F}_A = \frac{\partial \bar{\mathbf{x}}}{\partial \xi^\alpha}\sigma^{\alpha\beta}\frac{\partial N_A}{\partial \xi^\beta} A^e \tag{30}$$

$$^{B,e}\mathbf{F}_A = \sigma_B^{\alpha\beta}\frac{\partial^2 \psi_A}{\partial \xi^\alpha \partial \xi^\beta}\mathbf{n}\, A^e \tag{31}$$

$$^e M_{A'} = \sigma_B^{\alpha\beta}\frac{\partial^2 \phi_{A'}}{\partial \xi^\alpha \partial \xi^\beta} A^e \tag{32}$$

whereas the external equivalent nodal forces can be explicitly derived as,

$$^e\mathbf{R}_A = \frac{1}{3}P\,\mathbf{n}\,A^e \tag{33}$$

139

where P is the applied loading pressure. Finally, collecting internal and external force terms, the element equilibrium equations, for a constant membrane stress, constant moment Kirchhoff shell element emerges as,

$$^e\mathbf{F}_A = {}^{M,e}\mathbf{F}_A + {}^{B,e}\mathbf{F}_A = {}^e\mathbf{R}_A$$
$$^e M = 0 \tag{34}$$

Now representing the global internal and external equivalent forces as, $\mathbf{F}$ and $\mathbf{R}$ respectively, the equilibrium equations becomes

$$\mathbf{F}(\hat{\sigma}[\mu(\bar{\mathbf{x}}, \bar{\mathbf{v}}), \bar{\mathbf{x}}, \bar{\mathbf{v}}], \bar{\mathbf{x}}) = \mathbf{R}(\bar{\mathbf{x}}, t) \tag{35}$$

Since all the quantities are dependent on the solution variable velocity $\bar{\mathbf{v}}$, residual force vector $\mathbf{T}$ can be expressed as,

$$\mathbf{T}(\bar{\mathbf{v}}) = \mathbf{R}(\bar{\mathbf{v}}) - \mathbf{F}(\bar{\mathbf{v}}) \tag{36}$$

Now expanding the residual force vector $\mathbf{T}$ using Taylor's series and considering only the linear terms, the final finite element equation for solution can be simply be written as [12],

$$\mathbf{K}\Delta\bar{\mathbf{v}} = \mathbf{T} = \mathbf{R} - \mathbf{F} \tag{37}$$

The stiffness $\mathbf{K}$ can be represented in matrix form and consists of the following stiffness matrices [12].

1. Membrane: materially linear stiffness matrix

$$K^{ij}_{Mml,AB} = 2\mu^M\, h\, \frac{\partial \bar{x}^i}{\partial \xi^\alpha} \frac{\partial N_A}{\partial \xi^\beta} G^{\alpha\beta\gamma\delta} \frac{\partial \bar{x}^j}{\partial \xi^\gamma} \frac{\partial N_B}{\partial \xi^\delta}\, A^e \tag{38}$$

2. Membrane: materially nonlinear stiffness matrix

$$K^{ij}_{Mmnl,AB} = \frac{(m-1)}{3h\mu\bar{e}^2} \frac{\partial \bar{x}^i}{\partial \xi^\alpha} \frac{\partial N_A}{\partial \xi^\beta} \sigma_M^{\alpha\beta} \sigma_M^{\gamma\delta} \frac{\partial \bar{x}^j}{\partial \xi^\gamma} \frac{\partial N_B}{\partial \xi^\delta}\, A^e \tag{39}$$

3. Membrane: geometrically nonlinear stiffness matrix

$$K^{ij}_{M\sigma,AB} = \delta^{ij} \frac{\partial N_A}{\partial \xi^\alpha} \sigma_M^{\alpha\beta} \frac{\partial N_B}{\partial \xi^\beta}\, A^e\, \Delta t \tag{40}$$

4. Bending: stiffness matrices

$$K^{ij}_{B1,AB} = 2\bar{\mu}^B \frac{h^3}{12} \frac{\partial^2 \psi_A}{\partial \xi^\alpha \partial \xi^\beta} G^{\alpha\beta\gamma\delta} \frac{\partial^2 \psi_B}{\partial \xi^\gamma \partial \xi^\delta} n^i n^j\, A^e \tag{41}$$

$$K^i_{B1,AB'} = 2\bar{\mu}^B \frac{h^3}{12} \frac{\partial^2 \psi_A}{\partial \xi^\alpha \partial \xi^\beta} G^{\alpha\beta\gamma\delta} \frac{\partial^2 \phi_{B'}}{\partial \xi^\gamma \partial \xi^\delta} n^i\, A^e \tag{42}$$

$$K_{B3,A'B'} = 2\bar{\mu}^B \frac{h^3}{12} \frac{\partial^2 \phi_{A'}}{\partial \xi^\alpha \partial \xi^\beta} G^{\alpha\beta\gamma\delta} \frac{\partial^2 \phi_{B'}}{\partial \xi^\gamma \partial \xi^\delta}\, A^e \tag{43}$$

<u>Solution Procedure</u>

1. Set initial conditions: $t_0 = 0$, $\bar{x} = 0$, $\bar{v} = 0$.
2. An initial stress distribution corresponding to the optimum strain-rate is assumed in the sheet. This initial stress value enables the evaluation of the stiffness matrices in the undeformed state. A user specified pressure is applied to the sheet and the sheet is allowed to deform. Since the sheet is initially flat at time $t_0 = 0$, the sheet is not capable of carrying membrane stresses, the deformation occurs purely due to bending contributions. For the second step the in-plane membrane stress is assumed to be equal to the maximum bending stress solely for the purpose of making the sheet capable of handling membrane loads.
3. The second load step is also executed using a user specified pressure load. The velocities are solved for. Membrane and bending strain-rates are first computed followed by the stresses. These values are now used for the computation of stiffness matrices for the next load step.
4. The stiffness matrices are computed and assembled.
5. The solution is obtained after Newton-Raphson iterations are performed for convergence. A convergence criteria based on both Force norm and Velocity norm [12] are used to check for convergence.
6. The maximum strain-rate in the domain is computed and compared with the optimum strain-rate. If the maximum strain-rate exceeds or falls below the tolerance bounds of $\pm 5\%$ of the optimum strain-rate, the loading pressure is corrected using the correction scheme. The geometry and other quantities are reset to corresponding values in the previous time step and steps 4 through 6 are repeated. If the maximum strain-rate falls within the tolerance bounds the program continues to the next step.
8. Contact algorithm is used to check for contact. If contact is detected, the exact point(s) of contact and depth of penetration are computed. With the depth of penetration the velocity required for the compatibility load step is computed and applied to the solution scheme. The new equilibrium configuration is now evaluated.
9. Now the pressure control algorithm maintains the optimum strain-rate and the pressure for the next load step is predicted.
10. Geometry and thickness of each element is updated. The thickness is computed by maintaining constancy of volume and hence

$$h(t_n) = h(t_0)\frac{A(t_0)}{A(t_n)} \tag{44}$$

where h is the thickness and A is the area of the element. The strain-rates, stresses and equivalent strains are computed for each element and stored appropriately. The various indices are reset for the next time step. If the maximum number of load steps is not exceeded proceed to step 4 for the next load increment.

<u>Contact Algorithm</u>

A three dimensional contact algorithm has been developed similar to the two dimensional contact procedure explained in Reference [13]. Essentially the contact problem is solved by imposing geometric constraints on the configuration at which the compatibility conditions are violated (*pseudo- equilibrium configuration*). Such a procedure is easily adapted to a displacement formulation but with modifications this concept is easily incorporated in the velocity formulation described above.

In contact problems, nonlinearity in the formulation arises from both the variation of the area of the contact surfaces and also from the effects of friction. For simplicity, consider the case of Hertzian contact where successive nodes come into contact as the

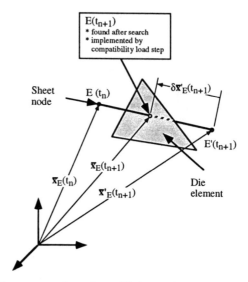

Figure 3. The concept of pseudo equilibrium and compatibility load step

load increases. Assuming that there is no contact at t_n the finite element equations can be represented as

$$[K]\{\Delta v\} = \{R\} - \{F\} = \{T\} \qquad (45)$$

where $[K]$ is the global stiffness matrix, $\{\Delta v\}$ is the incremental global velocity vector used in the Newton-Raphson iterative solution scheme, and $\{R\}$ is the applied force vector, while $\{F\}$ is the internal force vector. Let $\{T\}$ be the residual force vector.

Let $\bar{\mathbf{x}}_E(t_n)$ be the position vector of node E on the deforming sheet at time t_n (Figure 3). The body is at equilibrium but at the next load step the new position is given by $\bar{\mathbf{x}}'_E(t_{n+1})$ which is not at equilibrium since it has violated the contact conditions. Let the configuration at this pseudo-equilibrium state be $\Omega'(t_{n+1})$.

Now there exists a configuration $\bar{\mathbf{x}}_E(t_{n+1})$ such that the node E is just in contact with the die element and it satisfies the kinematic contact conditions. To obtain this configuration let us first assume that the displacement $\Delta\bar{\mathbf{x}}'_E(t_{n+1})$ from $\bar{\mathbf{x}}_E(t_n)$ to $\bar{\mathbf{x}}'_E(t_{n+1})$ be linearly varying and thus described as

$$\Delta\bar{\mathbf{x}}'_E(t_{n+1}) = \bar{\mathbf{x}}'_E(t_{n+1}) - \bar{\mathbf{x}}_E(t_n) \qquad (46)$$

The next step is to exactly compute the location of the node E on the die element $\bar{\mathbf{x}}_E(t_{n+1})$. A search strategy assuming that the displacements in each time step are linear, is employed to compute the exact location(s) of the contact node(s). Details of this procedure are given in reference [12].

The depth of penetration is given by

$$\delta\bar{\mathbf{x}}'_E(t_{n+1}) = \bar{\mathbf{x}}_E(t_{n+1}) - \bar{\mathbf{x}}'_E(t_{n+1}) \qquad (47)$$

If $\delta\bar{\mathbf{x}}'_E = 0$ then the node E is exactly in contact with the die. This concept was first introduced by Chandra [13] and later used in the formulation for superplastic sheet forming processes [2,3].

142

In the velocity formulation, the positions of all global nodes are determined at the end of each equilibrium load step as

$$\Delta \bar{\mathbf{x}}(t_{n+1}) = [\zeta \bar{\mathbf{v}}(t_n) + (1 - \zeta)\bar{\mathbf{v}}(t_{n+1})] \Delta t_n$$
$$\bar{\mathbf{x}}(t_{n+1}) = \bar{\mathbf{x}}(t_n) + \Delta \bar{\mathbf{x}}(t_{n+1}) \tag{48}$$

where $\bar{\mathbf{v}}(t_n)$ and $\bar{\mathbf{v}}(t_{n+1})$ are the global velocity vectors at times t_n and t_{n+1} respectively. Δt_n denotes the time increment at a time t_n and ζ is the update factor which varies from 0 to 1 depending upon the updating procedure adapted. Its value is set to 0.5 in this formulation to signify the Crank-Nicholson method of updating.

The compatibility load step has to be performed on the pseudo-equilibrium configuration to get the actual equilibrium configuration at time t_{n+1}. For this a displacement equal to the magnitude of penetration $\delta \bar{\mathbf{x}}'(t_{n+1})$ has to be applied in the opposite direction to the pseudo-equilibrium configuration. The independent variable in the finite element formulation is velocity and not displacements. Therefore, to apply a certain displacement, the corresponding incremental velocity and time increment have to be found. For a certain time increment equal to the user specified value, the incremental velocity field can be calculated as,

$$\Delta \bar{\mathbf{v}}(t_{n+1}) = \bar{\mathbf{v}} - \bar{\mathbf{v}}'(t_{n+1}) = \frac{1}{\theta \Delta t_{n+1}} \left[\delta \bar{\mathbf{x}}'(t_{n+1}) - \bar{\mathbf{v}}'(t_{n+1}) \times \Delta t_{n+1} \right] \tag{49}$$

This incremental velocity is directly applied.

Thus, components of the velocity vector at a typical penetrating node E are determined which will produce the desired displacements in the time interval to annul the contact overlap. The velocities can be imposed at the degree of freedom i using the direct method suggested in reference [12].

Pressure predictor-corrector method

It is necessary to monitor and control the maximum strain-rate in the sheet so that it does not exceed the optimum strain-rate. Such stringent control is required to maintain optimum superplastic condition during the forming process. The first step in the implementation of the pressure control algorithm is the determination of the maximum value of strain-rate and its actual location in the domain. This maximum strain-rate is then compared with the specified optimum strain-rate to predict the new pressure for the next load step. To derive a direct relationship between the loading pressure and the kinematic quantity strain-rate requires certain assumptions. Since superplastic sheet forming is mostly a membrane stretching process, it can be treated as a deforming membrane. From membrane theory, it is known that the pressure is directly proportional to the stress. Hence using the uniaxial power-law model ($\bar{\sigma} = K\dot{\bar{\epsilon}}^m$) for superplastic materials, the ratio of pressures can be represented as,

$$\frac{p^{t+\Delta t}}{p^t} \simeq \frac{\bar{\sigma}^{t+\Delta t}}{\bar{\sigma}^t} = \left(\frac{\dot{\bar{\epsilon}}^{t+\Delta t}}{\dot{\bar{\epsilon}}^t} \right)^m \tag{50}$$

where the right superscript indicate that the quantities are at specified time. With the knowledge of all the quantities at time t, it is required to predict the pressure at time $t + \Delta t$, that would produce a maximum strain-rate in the domain close to the required optimum strain-rate $\dot{\bar{\epsilon}}_{opt}$. The pressure in the previous step p^t and the strain-rate at that step, $\dot{\bar{\epsilon}}^t$ are known, and by setting $\dot{\bar{\epsilon}}^{t+\Delta t}$ equal to the optimum strain-rate, $\dot{\bar{\epsilon}}_{opt}$, equation (50) can be rewritten to predict the new loading pressure for the next load step, $p^{t+\Delta t}$. To avoid numerical stabilities, a weighted averaging scheme was used which includes the effect of pressure at time $t - \Delta t$ also. Hence the pressure predicting equation can be expressed as,

$$p^{t+\Delta t} = \nu \left(\frac{\dot{\bar{\epsilon}}_{\text{opt}}}{\dot{\bar{\epsilon}}^{t}} \right)^{m} p^{t} \ + \ (1 - \nu) \left(\frac{\dot{\bar{\epsilon}}_{\text{opt}}}{\dot{\bar{\epsilon}}^{t-\Delta t}} \right)^{m} p^{t-\Delta t} \tag{51}$$

where the weighting factor ν varies from 0 to 1, but is assumed to be 0.5 in this implementation. A new pressure is computed for every new load step. Then the maximum strain-rate in the domain is located and compared with the optimum strain-rate. If the maximum strain-rate is within the user specified tolerance level of the optimum strain-rate then procedure continues with the prediction of new pressures for the subsequent steps. If the maximum strain-rate lies out of the tolerance range, then the pressure needs to be corrected for the same load step. Hence a correction scheme based on the same method as predictor scheme can be devised as,

$$^{(i+1)}p^{t} = {}^{(i)}p^{t} + \theta \times \left[\left(\frac{\dot{\bar{\epsilon}}}{{}^{(i)}\dot{\bar{\epsilon}}^{t}} \right) - 1 \right] {}^{(i)}p^{t} \tag{52}$$

where the left superscript denotes the iterative step number for the pressure correction procedure. Here the weighting factor θ is not the same as ν in the predicting scheme. θ is set to 0.5 during the initial stages until an instance of contact. When contact occurs θ is set to 1 and hence the effects of the previous steps are ignored in the correction scheme making it easier to avoid numerical instabilities, thus increasing the chance of convergence. The iterations are carried out until $^{(i)}\dot{\bar{\epsilon}}^{t}$ is within the allowable range of $\dot{\bar{\epsilon}}_{\text{opt}}$. A more detailed description of the development of this algorithm is provided in References [12,14].

RESULTS AND DISCUSSIONS

A code has been developed based on the formulation and solution procedure explained in the previous chapters. The program called SPASM3D (SuperPlastic Analysis of Sheet Metals - 3D) is coded in FORTRAN77 and was initially implemented in a VAX/VMS computing environment. Once the contact and pressure prediction-correction algorithms were developed and added to the code, modifications to the code were made so that it could be easily adapted to execute in machines with UNIX operating systems. Currently, the existing program can be executed without modifications on VAX/VMS, Sun/UNIX and Silicon Graphics/IRIX Workstations and also on the CRAY-YMP running UNICOS. Even though it does not need a supercomputer the program was mostly executed on the CRAY-YMP for convenience.

Generally, demonstration parts such as cups, cones, long rectangular boxes and domes, are superplastically formed to study the characteristics of new superplastic materials and also to prove the viability of SPF as an economical manufacturing method.

Three different cases analysed using SPASM3D are discussed here. At first, the code is validated using the simulation of a 60^0 angle cone. The second example involves superplastic forming of a long rectangular pan with bottom stiffeners. The results were compared with that of two dimensional continuum model (SPASM2D) [2]. This simulation proved to be a good test for the contact algorithm as well as the application of dual strain-rate criteria for handling dynamically recrystallizing materials. The last example demonstrates the ability of the code to model a realistic three dimensional component. The purpose of this example is to extend the analytical capability of the code to design the superplastic forming of a practical component. Parameters for the process design, such as pressure-time cycle and final thickness distribution are obtained. Discussions to assess the suitability of the component for superplastic forming are provided.

<u>Cone</u>

When superplastic sheets are formed into cones with constant or varying cone angles, the primary goal of this test is to verify the formability of superplastic materials under balanced biaxial conditions. In this example a cone with a constant angle of 60° is considered. The superplastic material Al 7475 (ALCOA) was considered, with $K = 338290.0$ psi.sec^{-m}, the strain-rate sensitivity $m = 0.6324$ and with optimum strain-rate $\dot\epsilon_{opt} = 0.0007$ sec^{-1}. The geometric details of the cone are shown in Figure 4. The die discretized using 30 die elements (Figure 5).

When using the three dimensional model for this problem only a quarter of the cone was considered due to symmetry. The condition of axisymmetry is achieved in the problem through the application of boundary conditions. The quarter section of the circular sheet of initial thickness 0.09 in. is discretized using 160 elements, with 353 nodes and 547 degrees of freedom (Figure 6). The deformation stages of the cone is given by Figure 8.8. During deformation of the cone, it is observed that the circular sheet (one quarter) deforms with a more or less uniform strain-rate throughout the sheet, in the initial stages. Once contact is established with the die walls, the area of the free forming region gradually reduces. The control point for pressure prediction is always located in the free forming region. A slight variation in thickness was observed when traversing radially from the pole of the free forming sheet to the edge where the sheet is in contact with the die. Since a sticking contact condition is assumed there is no deformation of the elements that are already in contact. The pressure-time profile for the superplastic forming of this cone is shown in Figure 7.

The thickness profile of the final formed part (Figure 8) is compared with the experimental data that was available from the superplastic cone provided by LTV Aerospace. Excellent agreement is observed between the numerical and experimental results. It takes 55 min. 12 secs. on a CRAY-YMP for the complete forming of the cone. The total forming time predicted is 54 min. 40 secs.

<u>Long Rectangular Pan with Bottom Stiffeners</u>

This specific example is chosen to illustrate the applicability of the dual strain-rate criteria for dynamically recrystallizing materials. This has been previously illustrated by SPASM2D using continuum elements [15]. The assumption of plane strain in the length direction is imposed through boundary conditions applied on a thin strip of sheet (Figure 9). The sheet strip is discretized into 144 elements with 365 nodes. Al-Li 2090 alloys described by the constitutive relationship $\sigma = K\dot\epsilon^m$ is used in this case, with $K = 594414$ psi sec^{-m} and $m = 0.8$. The initial thickness of the sheet is assumed to be 0.08 in. The die is constructed using 22 three noded triangular elements. The geometry of the die mesh is given by Figure 10.

Dual strain-rate criterion is applied by specifying a strain-rate of 0.002 sec^{-1} till the transition strain (ϵ_T) of 0.25, and the optimum strain-rate of 0.0002 sec^{-1} beyond the transition strain. Figure 11 shows the various stages of deformation for the pan. The convex surfaces in the bottom depicting bottom stiffeners, proves to be a severe test for the contact algorithm. The comparison of pressure-time curves between the 2D continuum model (SPASM2D) and 3D thin shell model (SPASM3D), for the dual strain-rate case is shown in Figure 12. There is a good correlation between the profiles except in the final stages of forming. The final stages correspond to the forming of the sheet into die corners. The difference in discretization of the sheet in the 2D and 3D cases have a strong influence on the pressure-time curve. The final thickness profiles of the 2D and 3D models are compared for the dual strain-rate case (Figure 13). Reasonable agreement in the thickness profile is obtained. The fine discretization in the 3D model has enabled it to predict even minor variations in thicknesses between adjacent elements.

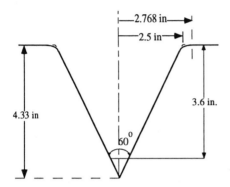

Figure 4. Die details of the 60° angle cone

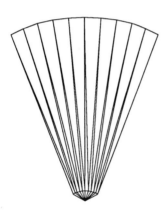

Figure 5. Finite element mesh of the die
for the SPF of the cone (quarter)

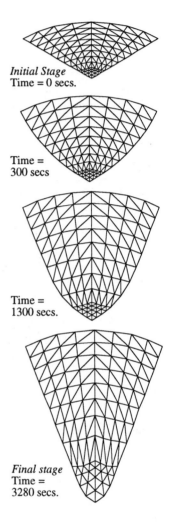

Initial Stage
Time = 0 secs.

Time =
300 secs

Time =
1300 secs.

Final stage
Time =
3280 secs.

Figure 6. Deformation stages in
the SPF of the 60° angle cone

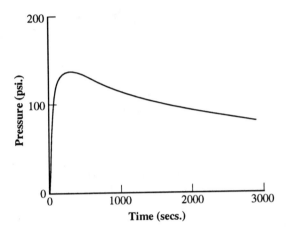

Figure 7. The pressure-time profile for the SPF of the 60° angle cone

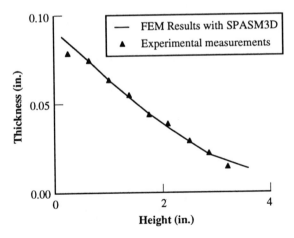

Figure 8. Experimental comparison of the final thickness
distribution vs. height of the cone

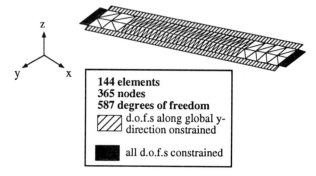

144 elements
365 nodes
587 degrees of freedom
▨ d.o.f.s along global y-
direction onstrained

■ all d.o.f.s constrained

Figure 9. The initial mesh for the three dimensional strip with the applied
boundary conditions to force plane strain

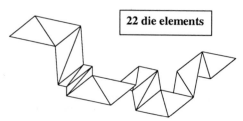

22 die elements

Figure 10. The finite element mesh for the die representing the
long rectangular pan with bottom stiffeners

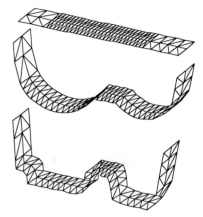

Figure 11. Deformation stages of the three dimensional strip representing
the SPF of complex long rectangular pan

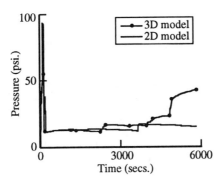

Figure 12. Pressure-time profile for the SPF of the complex
long rectangular pan by the 3D strip

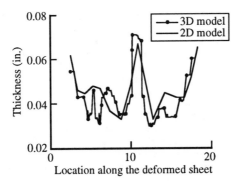

Figure 13. Comparison of the final thickness distribution predicted by the
2D and 3D models for the final formed shape

There is a savings of 17% in total forming time, predicted by both the 2D and 3D models with use of a dual strain-rate criterion instead of a single strain-rate control. The important observation was that the final thickness distribution for the single and dual strain-rate criterion remained virtually unchanged.

Complex Pan

The final example involves the design and analysis of a typical industrial superplastic component. In fact, the component is a candidate for potential use in a defense aircraft.

This example is similar to the *drape forming technique* even though it is seen as pan with convex shapes on the die bottom. For the sake of simulation, the problem can be treated as the forming of a complex pan. Details of the final component are given in Figure 14. The shaded portions in the figure have been neglected to achieve symmetry for ease of simulation. Thus only one quarter of the pan is modeled due to symmetry. The die is constructed using 54 die elements as shown in Figure 15. The sheet is discretized using 358 elements and 771 nodes with 1185 degrees of freedom (Figure 16a). The sheet is Al 7475 (ALCOA) described using $\sigma = K\dot{\epsilon}^m$ with $K = 338290$ psi sec^{-m} and $m = 0.6324$. The initial thickness of the sheet is 0.09 in. The various stages of deformation of the complex pan are shown in Figures 16a-d. The forming pressure-time cycle to maintain an optimum strain-rate of 0.0007 sec^{-1} is given in Figure 17. The band contours of the thickness distribution in the final component is given by Figure 18.

It is found in this simulation that more discretization is required to capture the draping process over the convex projections in the bottom. Moreover, since there are too many corners it is now important to capture the effects of shear and bending during corner forming, more than is capable by this thin shell element used in this formulation. Since this is a membrane dominated element it assumes that the deformations stop once the element nodes contact the die surface and hence prevents further deformation.

SUMMARY AND CONCLUSIONS

Superplastic forming is widely accepted by the aerospace industry as an advanced manufacturing technique to produce complex components in one forming operation. It is known that superplastic forming, despite its advantages in forming complex components requires stringent control of process parameters, and hence is not an inexpensive manufacturing method. This is in addition to the requirement to form complex shaped components successfully in the shortest process time. The finite element method proves to be a viable approach to handle this highly complex problem of modeling the superplastic sheet forming process. The examples presented in this paper show that a simple Kirchhoff thin shell formulation is sufficient to model SPF in order to provide the process designer with data on the forming history of the shape and thickness distribution. The iterative strain-rate control and pressure prediction procedure is found to be effective and simple. The final example shows the capabilities of the code developed on this formulation in predicting the process parameters for the SPF of a typical industrial component.

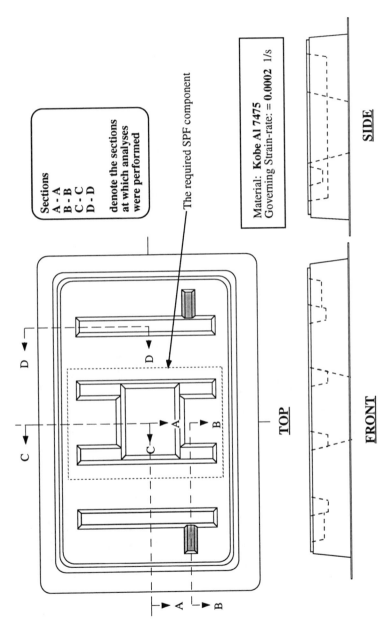

Sections
A - A
B - B
C - C
D - D

denote the sections
at which analyses
were performed

The required SPF component

Material: **Kobe Al 7475**
Governing Strain-rate: = **0.0002** 1/s

TOP

FRONT

SIDE

Figure 14. Schematic of the complex pan and the required SPF component

151

54 die elements

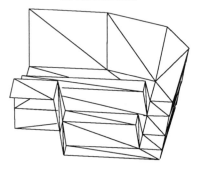

Figure 15. Finite element mesh of one quarter of the die
for the SPF of complex pan

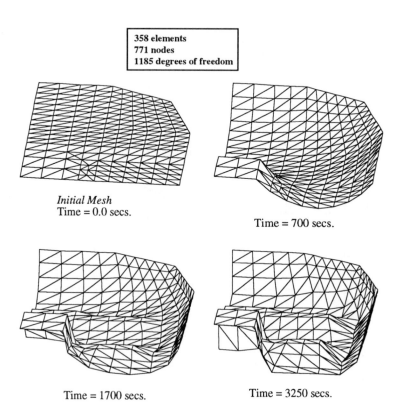

Figure 16. Deformation stages of the complex pan

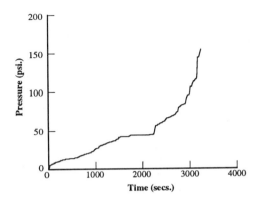

Figure 17. The pressure-time profile for the SPF of complex pan

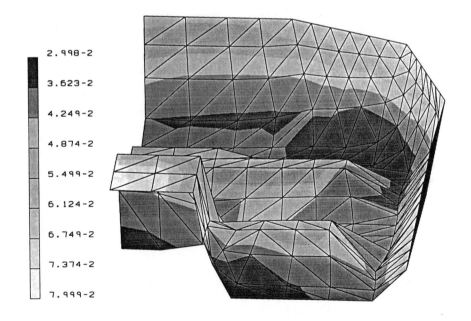

Figure 18. Thickness distribution in the fully formed SPF component

REFERENCES

1. K. A. Padmanabhan and G. J. Davies, Superplasticity (Berlin: Springer-Verlag, 1980).

2. N. Chandra, "Analysis of superplastic metal forming by a finite element method," Int. J. Numer. Methods Eng., 26 (1988), 1925-1944.

3. N. Chandra and S. C. Rama, "Application of finite element method to the design of superplastic forming processes," ASME Journal of Engineering for Industry, 114 (1992), 452-458.

4. N. Chandra, S. C. Rama and R. E. Goforth, "Process modeling of superplastic forming processes using four different computational methods," Superplasticity in Advanced Materials, ed. S. Hori, M. Tokizane and N. Furushiro, (Japan: The Japan Society for Research on Superplasticity, 1991), 837-844.

5. S. C. Rama and N. Chandra, "Finite element analysis of superplastic forming processes using continuum and thin shell formulations," Advances in Finite Deformation Problems in Materials Processing and Structures, AMD-Vol. 125, ed. N. Chandra and J. N. Reddy, (New York, NY: ASME Publications, 1991), 131-145.

6. M. Bellet and J. L. Chenot, "Numerical modeling of 3-dimensional superplastic forming of titanium sheet," Proc. of Int. Conference on Titanium Products and Applications, Volume 2, (Sophia Antipolis, France: 1986), 1175-1184.

7. M. Bellet, E. Massoni and J. L. Chenot, "A viscoplastic membrane formulation for 3-dimensional analysis of superplastic forming of thin sheet," Proc. Int. Conf. on Computational Plasticity Model, Software and Applications, Part 2, (Sophia Antipolis, France: 1987), 917-926.

8. M. Bellet and J. L. Chenot, "Numerical modeling of thin sheet superplastic forming," Numiform '89, ed. Thompson et al., (Brookfield, VT: Balkema Publishers, 1989), 401-406.

9. J. Bonet, R. D. Wood and O. C. Zienkiewicz, "Finite element analysis of thin sheet superplastic forming," Superplasticity and Superplastic Forming, ed. C. H. Hamilton et al., (Warrendale, PA: The Minerals, Metals and Materials Society, 1988), 291-295.

10. J. Bonet, R. D. Wood and A. H. S. Wargadipura, "Simulation of the superplastic forming of thin sheet components using the finite element method," Numiform '89, ed. E. G. Thompson et al., (Brookfield, VT: Balkema Publishers, 1989), 85-93.

11. J. Bonet, R. D. Wood and A. H. S. Wargadipura, "Numerical simulation of the superplastic forming of thin sheet components using the finite element method," Int. J. Numer. Meth. Eng., 30 (1990), 1719-1737.

12. S. C. Rama, "Finite element analysis and design of 3-dimensional superplastic sheet forming processes," (Ph. D. Dissertation, Texas A&M University, 1992).

13. N. Chandra, W. E. Haisler and R. E. Goforth, "A finite element solution method for contact problems with friction," Int. J. Numer. Methods Eng., 24 (1987), 477-495.

14. S. C. Rama and N. Chandra, "Development of a pressure prediction method for superplastic forming processes," Int. J. Non-Linear Mech., 26 (5) (1991), 711-725.

15. S. C. Rama, N. Chandra and R. E. Goforth, "Computational process modeling of dynamically recrystallizing superplastic materials," ed. J. L. Chenot et al., Numiform '92, (Rotterdam, Netherlands: Balkema Publishers, 1992), 867-872.

THE EFFECTS OF PROCESSING PARAMETERS ON THE POST-SPF

MICROSTRUCTURE AND MECHANICAL PROPERTIES OF WELDALITE™ 049

P. J. Smith-Hartley
Martin Marietta Manned Space Systems, New Orleans, LA 70189

K. S. Kumar and S. A. Brown
Martin Marietta Laboratories, Baltimore, MD 21227

Abstract

In this study, an alloy belonging to a class of Al-Cu-Li alloys under the trade name
Weldalite™ 049, was examined for its potential to be superplastically formed into a slosh
baffle panel, a component in the external fuel tank (ET) of the Space Shuttle. Various
forming routes were evaluated. Secondary heat treatment schedules were examined for each
forming route to identify overall optimal processing conditions to successfully fabricate the
panel with desirable mechanical properties. Microstructural characterization and tensile
property evaluation in various orientations were instrumental in identifying the appropriate
processing parameters.

Advances in Superplasticity and Superplastic Forming
Edited by N. Chandra, H. Garmestani, R.E. Goforth
The Minerals, Metals & Materials Society, 1993

Introduction

The utilization of aluminum-lithium alloys (Al-Li) offers a potential to increase payload capability of launch vehicles in the aerospace industry through the reduction of vehicle weight. Advanced manufacturing techniques that are capable of producing near-net shape parts from these alloys will aid in reducing the number of processing and assembly steps, thereby reducing launch system costs. One such processing route is Superplastic Forming (SPF). Superplastic formability of aluminum alloys has been extensively examined (1, 2). Recently, attention has been given to SPF of Al-Li alloys (3, 4) due to the attractive specific strength and modulus that they exhibit. An inherent disadvantage of Al-Cu-Li alloys is that they typically require a stretching operation prior to age-hardening to provide the dislocations that are necessary for nucleating the T_1 (Al_2CuLi) precipitates. These T_1 precipitates are primarily responsible for the superior strength that is typically obtained in these alloys. This characteristic is a drawback because it is not possible to stretch superplastically formed parts and therefore, the aging response of the unstretched material is of primary concern.

Recently, a family of Al-Cu-Li alloys containing additions of Ag, Mg, and Zr has been developed under the trade name Weldalite™ 049 (5). These alloys have been shown to exhibit an unusually strong aging response without the stretching operation; post-aging properties (i.e., T6 condition) have been shown to be superior to other commercial alloys that have even been previously stretched (i.e., T8 condition). If these alloys are amenable to SPF, the near-net shaped part can then be solution-treated and aged to attain attractive properties. Mahon and Ricks (6) have recently reported on the superplastic formability of Weldalite™ 049 and have shown that these alloys exhibit favorable strain rate sensitivity values (resistant to necking) at strain rate levels that are practical for commercial operations. Further, they showed that the optimum temperature for superplasticity is similar to the solution treatment temperature so that it may be possible to combine these two processing steps. The positive influence of an applied hydrostatic back pressure during SPF on elongation prior to failure was also confirmed.

In this study, the effect of superplastic forming parameters and subsequent aging conditions on the microstructure and mechanical properties of Weldalite™ 049 has been examined for a specific part of the external fuel tank (ET) of the Space Shuttle and an optimal processing route for this component is recommended.

Experimental Procedure

The slosh baffle (Figure 1), a component in the liquid oxygen tank of the ET of the Space Shuttle was identified as a candidate for SPF. The SPF process offers the attractive potential to decrease the number of parts (fasteners and stiffeners) currently required to assemble this

Figure 1 -   Photograph of the slosh baffle panel.

Table I - Processing Routes for the SPF Baffles.

Nomenclature	Process Definition
SPF/FAC	SPF @ 504°C + Force Air Cool to Room Temperature
SPF/WQ	SPF @ 504°C + Water Quench to Room Temperature
SPF/SHT/WQ	SPF @ 495°C + SHT @ 504°C +Water Quench to Room Temperature

Table II - Forming Parameters used to Fabricate the Slosh Baffles.

Gage (mm)	Forming Temp Start, Finish (°C)	Back Pressure (MPa)	Forming Time (min)	Cooling Condition
1.0	505, 505	2.48	25	WQ
1.0	509, 505	2.41	25	WQ
1.0	503, 503	2.41	25	WQ
1.0	504, 503	2.41	25	FAC
1.0	504, 505	2.41	25	FAC
1.0	506, 506	2.41	25	FAC
1.0	506, 505	1.38	25	FAC
1.0	493, 495	2.41	25	FAC
1.0	494, 495	2.41	25	FAC
1.0	495, 496	2.41	25	FAC
1.0	495, 495	1.38	25	N/A
1.0	493, 495	1.38	25	FAC
1.0	495, 490	2.41	19	FAC
1.0	489, 490	2.41	25	FAC
1.0	489, 488	2.41	25	FAC

component and hence, significant cost reductions. The composition of the alloy used in this study contained (in wt%) Al-4.6Cu-1.14Li-0.4Ag-0.4Mg-0.14Zr and was obtained in sheet form (2.2 mm) from Reynolds Metals Company. A processing technique, proprietary to Reynolds, was used in producing the sheet with a microstructure conducive to SPF. Specimens from this sheet were solution treated at 504°C/1h and water quenched to room temperature. Aging curves at 160°C, 171°C, and 182°C were obtained and near peak age conditions were identified (i.e. time/temperature combination). Tensile tests were conducted on specimens thus heat treated. Specimens were obtained in the longitudinal (L), long transverse (LT), and 45° orientation to the rolling direction. Yield strength, ultimate tensile strength (UTS), and elongation were measured and documented. Fracture surfaces were characterized by scanning electron microscopy (SEM).

To accommodate redesign requirements of the slosh baffle, the 2.2 mm sheet was chemically milled to a thickness of 1 mm. The components were then fabricated with a forming pressure of 2.75 MPa and back pressures up to 2.42 MPa. Three different SPF processing routes (Table I) were selected. For the first two conditions, the SPF temperature was selected to be the same as the solution treatment temperature with the intention of eliminating the additional solution treatment step that was necessary in the third processing route. The cross section of the SPF part in each instance was optically examined to identify any variation that might exist in the microstructure across the thickness. Details of the actual forming parameters are provided in Table II. Aging curves at 160°C, 171°C, and 182°C were once again developed for the superplastically formed material. Optimal aging conditions were identified and tensile specimens machined from the part were thus heat treated and tested. Specimens were machined in the three previously described orientations (still with respect to the original rolling direction) to obtain optimal combinations of strength and ductility, and a minimum in anisotropy.

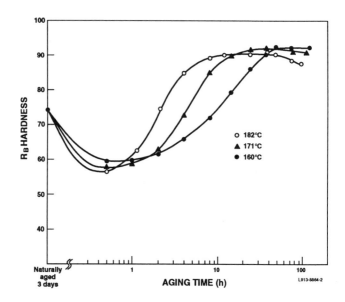

Figure 2 -   Aging response at 160°C, 171°C, and 182°C of the as-received rolled sheet
after it was solution treated at 504°C/1h and naturally aged for three days.

Results and Discussion

Characterization of the As-Received Sheet

Aging Response. Sections from the as-received sheet were solutionized at 504°C for one
hour. Following solution treatment, specimens were allowed to naturally age for three days
and then were artificially aged at 160°C, 171°C, and 182°C for various times. Hardness was
measured as a function of aging time at each temperature. A reversion occurs during the
early part of this aging response accompanied by a dramatic decrease in hardness (Figure 2).
Microstructural changes accompanying this reversion behavior in similar alloys have been
recently discussed (7). For further aging times, at all temperatures investigated, hardness
increases to a peak and the time to reach peak hardness decreases with increasing
temperature. From these aging curves (Figure 2), suitable combinations of time/temperature
corresponding to near-peak age condition were identified.

Mechanical Testing. Next, tensile specimens were machined from the rolled sheet in L,
LT, and 45° orientations. These were solution treated at 504°C/1h and water quenched to
room temperature. After three days of natural aging, specimens were artificially aged at
160°C/36h, 171°C/14h, or 182°C/8h to obtain near-peak age properties. Following these heat
treatments, room temperature testing was performed. The effect of aging parameters on the
yield strength, UTS and elongation are shown in Table III. The yield strength for specimens
in the L and LT orientations are higher following heat treatments at 171°C and 182°C relative
to 160°C without a significant ductility penalty. In general, the yield strength of the 45°
orientation specimens are lower than the L and LT orientation for a given heat treatment; the
ductility however is higher.

Table III - Mechanical Properties of the As-Received Weldalite™ 049 After Heat Treatment.

Aging Parameters (°C)	(h)	Orientation	UTS (MPa)	YS (MPa)	el (%)
160	36	L	606	545	11.0
		LT	582	516	17.4
		45°	517	470	17.8
171	14	L	628	581	10.0
		LT	613	577	10.0
		45°	544	505	16.5
182	8	L	615	563	9.3
		LT	607	568	9.5
		45°	532	492	16.9

Representative micrographs of fracture surfaces of the tensile specimens (Figure 3a-c) show a preponderance of intersubgranular failure although occasionally evidence of microvoid coalescence can be noted. The occurrence of subgrain boundary failure could be related to the precipitation of the T_1 phase at the subgrain boundaries or alternately to possible elemental Li segregation to the cores of dislocations that compose these boundaries, thereby embrittling these boundaries. These proposals are purely speculative and no effort was made in this investigation to identify the cause of this failure mode. It is also believed that the intrinsic yield strength of the material (i.e., underaged versus near-peak aged) influences the fracture mode to the extent that vastly underaged materials tend to fail by microvoid coalescence rather than in the predominantly intersubgranular mode, even though "sufficient" amounts of the T_1 phase have already precipitated both at subgrain boundaries and in the matrix.

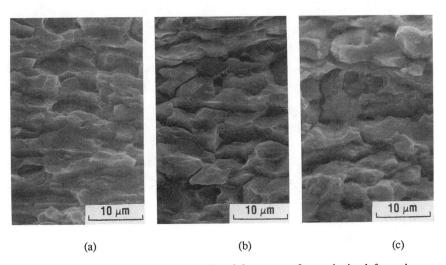

(a)　　　　　　　　　　(b)　　　　　　　　　　(c)

Figure 3 -　Representative micrographs of fracture surfaces obtained from the as-received, heat treated tensile specimens (504°C/1 h + water quench + 160°C/36 h): (a) L, (b) LT, and (c) 45° orientations.

Post-SPF Characterization

Microstructure. The microstructure of the slosh baffle panels formed via the three
processing routes were examined using an optical microscope. The extent of recrystallization
through the thickness of the part as well as the effect of quenching rate on precipitation
during cooling were determined. A higher degree of recrystallization and grain growth was
noted adjacent to the surface in all three cases (Figure 4a-c) although the situation was most
severe where the part was solution heat treated at 504°C after forming it at 495°C.

At the optical microscopy level, the amount of coarse precipitates present in both the
water quenched and air cooled conditions were about the same (Figure 5a,b) when the SPF
temperature was maintained at 504°C, although it is likely that forced air cooling would allow
more precipitation than water quenching. Such precipitates would perhaps be just below
optical microscope resolution level but fairly coarse on the transmission electron microscope
scale (i.e., 0.5-1 μm level). Interestingly, the amount of such coarse precipitates, resolvable
at the optical level appeared lower in the specimen taken from the slosh baffle panel that was
first formed at 495°C and then solution treated at 504°C and water quenched (Figure 5c).
This is likely related to the time at temperature that each component experienced because in
the first two processing routes (i.e., SPF at 504°C plus either water quench or forced air
cool), the time to form the part was typically 25 minutes whereas the third route provided an
exposure of 504°C solution treatment for 60 minutes.

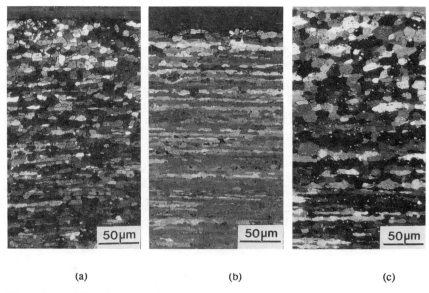

(a)   (b)   (c)

Figure 4 -   The effect of the SPF processing route on the microstructure of the formed
              part: (a) SPF at 504°C + water quench, (b) SPF at 504°C + force air cool, and
              (c) SPF at 495°C + solution heat treat at 504°C/1h + water quench.

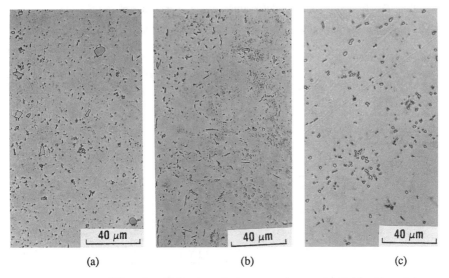

(a)	(b)	(c)

Figure 5 - The effect of the SPF processing route on the extent of "coarse" precipitates present in the formed part: (a) SPF at 504°C + water quench, (b) SPF at 504°C + force air cool, and (c) SPF at 495°C + solution heat treat at 504°C/1h + water quench.

Aging Response. Specimens from the slosh baffle panels obtained using the three processing routes were aged at 160°C, 171°C, and 182°C for various times to develop aging curves at each of these temperatures. The cross sectional thickness of these panels (1 mm) posed practical difficulties in obtaining reliable hardness measurements and thus Vickers hardness rather than Rockwell indentations was used. This resulted in a significant scatter in data, although hardness curves could still be generated with profiles analogous to those previously noted in the sheet material prior to forming but with some differences in the aging kinetics. From these aging curves, time/temperature combinations were identified to optimize strength and ductility in the longitudinal, long transverse and 45° orientations recognizing that in-plane anisotropy would require a compromise in mechanical properties.

Mechanical Properties. Tensile specimens from the L, LT, and 45° orientations (relative to the original rolling direction) were machined from the flat, bottom section of the slosh baffle panel. These were then aged at the three temperatures (160°C, 171°C, and 182°C) to an underaged and a near-peak aged condition (i.e., two different times at each temperature). Results of room temperature tensile tests on such specimens are given in Tables IV - VI.

For all the aging conditions examined, for all three SPF routes, the 45° orientation specimens exhibit the lowest strengths and therefore dictate the overall component failure conditions. While it may be possible to further age these parts in order to achieve additional strength enhancement in this orientation, this would likely occur at the expense of ductility in the other orientations. Therefore, it is evident that the optimal condition must take into account the ductility in the L and LT orientations and the strength in the 45° orientation.

Table IV - Post-SPF Mechanical Properties: Aging Temperature of 160°C.

Forming Process	Aging Time @ 160°C (h)	Orientation	UTS (MPa)	YS (MPa)	el (%)
SPF/FAC	36	L	451	321	12.0
		LT	477	362	13.0
		45°	441	360	16.0
	48	L	458	334	10.7
		LT	481	349	17.0
		45°	393	302	17.0
SPF/WQ	36	L	524	423	10.0
		LT	551	465	11.0
		45°	483	422	14.0
	48	L	553	473	5.0
		LT	566	489	8.5
		45°	515	459	9.0
SPF/SHT/WQ	36	L	567	484	9.0
		LT	558	476	10.0
		45°	461	402	13.0
	48	L	592	539	5.3
		LT	629	590	7.0
		45°	478	435	13.0

Table V - Post-SPF Mechanical Properties: Aging Temperature of 171°C.

Forming Process	Aging Time @ 171°C (h)	Orientation	UTS (MPa)	YS (MPa)	el (%)
SPF/FAC	14	L	468	342	10.0
		LT	465	329	14.0
		45°	424	344	19.0
	24	L	431	304	11.0
		LT	484	385	9.0
		45°	394	306	11.0
SPF/WQ	14	L	473	349	11.0
		LT	544	450	11.0
		45°	467	396	14.0
	24	L	519	438	5.3
		LT	495	397	8.0
		45°	505	403	9.5
SPF/SHT/WQ	14	L	569	486	8.0
		LT	583	519	10.0
		45°	484	441	12.0
	24	L	583	527	5.0
		LT	624	587	6.0
		45°	525	487	9.0

Table VI - Post-SPF Mechanical Properties: Aging Temperature of 182°C.

Forming Process	Aging Time @ 182°C (h)	Orientation	UTS (MPa)	YS (MPa)	el (%)
SPF/FAC	8	L	481	377	10.0
		LT	464	348	11.0
		45°	430	359	12.0
	24	L	422	305	9.0
		LT	483	399	8.0
		45°	441	371	11.0
SPF/WQ	8	L	513	416	8.0
		LT	527	454	7.0
		45°	463	398	11.0
	24	L	534	476	4.2
		LT	555	478	6.5
		45°	500	446	11.0
SPF/SHT/WQ	8	L	586	528	7.0
		LT	548	475	8.0
		45°	474	422	8.0
	24	L	436	336	8.0
		LT	565	494	7.0
		45°	389	301	11.5

An examination of the data in Tables IV - VI for the panels that were directly water quenched from the SPF temperature of 504°C reveals that the 160°C aging temperature is insufficient to obtain overall desired strength/ductility combinations. For example, 160°C/48h results in ductility in the L orientation being 5% whereas the yield strength in the 45° orientation is only ~460 MPa. The situation does not improve significantly by aging at 171°C or 182°C and in fact, the best combination obtained corresponds to 48h at 160°C. The aging response of the material subjected to the second processing route (SPF at 504°C/forced air cool) at all three temperatures for all aging times considered is disappointing. Yield strength in the 45° orientation does not exceed 400 MPa although sufficient ductility is realized in all three orientations. Whether this is a consequence of sluggish aging kinetics or essentially an upper limit in properties due to excessive precipitation during air cooling remains to be determined.

The panels that were formed at 495°C and then solution treated at 504°C and water quenched exhibit the best properties. The aging response of these panels can be appreciated by comparing the data in Tables IV - VI. The tensile properties after aging at 160°C/48h reveals a 45° orientation yield strength of 435 MPa and a ductility in the L orientation of 5.3%. When the aging conditions are changed to 171°C/24h, the ductility in the L orientation remains at 5.0%, whereas the yield strength in the 45° orientation increases to 487 MPa. Aging at 182°C results in properties that are inferior to the material aged at 171°C.

Based on these data, it is recommended that the baffle panels be formed at 495°C and then solution treated at 504°C/1h and water quenched. Subsequent aging at 171°C for 24h provides the best combinations of strength and ductility with reduced anisotropy.

Conclusions

A Weldalite™ 049 type alloy was superplastically formed using three different processing routes. The final part was a slosh baffle panel which is a component in the ET. This study has shown that the alloy is superplastically formable using temperature/strain rate/back pressure combinations that are practical. Surface recrystallization and grain growth was typical in this part for the conditions examined. Post-SPF properties identified an optimum forming approach that included a separate forming step at 495°C and a subsequent solution heat treatment at 504°C/1h. For a panel accordingly processed, optimal aging parameters were identified as 171°C/24h where the in-plane anisotropy in properties was the least and included yield strengths in the L, LT, and 45° orientations of 527 MPa, 587 MPa, and 487 MPa respectively and associated elongations of 5.0, 6.0, and 9.0%.

Acknowledgements

The authors are grateful to the Martin Marietta Research and Development program for sponsoring this work. We would like to acknowledge A.J. Barnes of Superform USA for the fabrication of the slosh baffle panels, Westmoreland Mechanical Test Laboratories for mechanical testing, and R. Greene and A. Cho of Reynolds Metals Company for supplying material and forming advice. Finally, we appreciate the inputs and support received from the Weldalite™ Technology Team of D. Bolstad, Bao-Tong Ma, J. Pickens, F. Heubaum and M. Dudley.

References

1. D. J. Lloyd and D. M. Moore: Superplastic Forming of Structural Alloys, ed: N. E. Paton and C. H. Hamilton, TMS-AIME, Warrendale, PA 1982, pp 147-172.

2. C. H. Hamilton, C. C. Bampton and N. E. Paton: ibid, pp 173-190.

3. J. Wadsworth, A. R. Pelton and R. E. Lewis, Metall. Trans., 1985, 16A (12), pp 2319-2332.

4. R. Grimes, W. S. Miller and R. G. Butler, J. de Phys. 48, C3, 1987, pp 239-249.

5. J. R. Pickens, F. H. Heubaum, T. J. Langan and L. S. Kramer: Proc. 5th Int. Conf. Al-Li Alloys, Williamsburg, VA, ed: T. H. Sanders and E. A. Starke, MCEP, Ltd., Birmingham, England, 1989, pp 1397-1414.

6. G. J. Mahon and R. A. Ricks, Scripta Metallurgica, 1991, 25 (2), pp 383-386.

7. F. W. Gayle, F. H. Heubaum, and J. R. Pickens: Proc. 5th Int. Conf. Al-Li Alloys, Williamsburg, VA, ed: T. H. Sanders and E. A. Starke, MCEP, Ltd., Birmingham, England, 1989, pp 79-84.

AUTHOR INDEX

Author Index